D1918555

Dancing with Change

Cultivating Healthy Organisations

Eric Lynn

a cultureQs book

German National Library bibliographic information.
This publication is recorded in the German National Bibliography.
Detailed bibliographic information is available via the following internet link:
https://dnb.d-nb.de.

Published by Eric Lynn, cultureQs
https://cultureqs.com/imprint/

Cover design: Tina Hanisch, missmilla design.
Printed by Scandanavian Book, c/o Druckhaus Nord GmbH,
Neustadt a. d. Aisch, Germany
Printed in Germany

ISBN 978-3-949056-00-0

"Mundus vult decipi:
The world wants to be deceived.
The truth is too complex and frightening:
the taste for the truth is an acquired taste that few acquire."

(Martin Buber, I and Thou)

Table of Contents

... Change ...
... and the Essence of Life ...

... Life is relational ...
... Independence is an illusion ... It is also real ...

... Change ... and ... Life ...

Welcome.

The focus of this book is Change. The focus is also Organisation and Society Health.
Actually, these two notions are so intricately intertwined that they are inseparable in the context of living communities.
While this may not be immediately obvious, hopefully, it will become so.

Convention suggests that these first pages be devoted to an introduction to the theme of this book.

Introductions are for beginnings.
As we are focusing on change, it makes little sense to speak of beginnings.
Why is this? ... Please read on ...

It is always healthy to Question norms and conventions.
We may or may not decide to change them.
Failure to Question them is potentially harmful to individuals and the societies in which we live.

We need to Question our narratives about change.
Indeed, it is essential that we change the way we think about change as well as the way we work with change.

Why?

Because our world is a mess.
We've got it wrong. Misunderstandings concerning the nature of change pervade life, and this is one of the primary reasons our world is a mess.

The mess includes … widespread violent conflict; millions of refugees fleeing conflict, poverty and exploitation; environmental destruction; the poverty of human ethics; pervasive corruption; the shallowness of politics; increasing levels of psychological stress; manipulation and wasted potential everywhere, including all corporations, institutions and organisations.
While the scale and perceived significance of these problems may be greater than in the past, none of them are in any way new.

The impact of these problems on health should be crystal clear.

What do they have to do with change?
Everything!

At the time of writing these words, it is May 2020, and throughout the world, people are trying to come to terms with the effects of the COVID-19 virus. However, notwithstanding the disruption that the virus is causing, our world was in a sad and unhealthy condition long before it enriched us with its presence.
"Enriched?", you may ask. Why, yes. COVID-19 has served to illuminate the relationships between so many of the factors causing the mess we have been creating for a long time.

Gregory Bateson, anthropologist, philosopher, biologist and much more besides, succinctly summarised the primary cause of our problems in the middle of the 20th century:

"The major problems in the world are the result of the difference between how nature works and how people think."

The problems we are confronted with today are infinitely more complex than they were during Gregory Bateson's lifetime.

Unfortunately, we, human beings, nowadays tend to listen far too rarely to those who offer us the learnings of this deep wisdom, wisdom that has actually been available to us for thousands of years, wisdom that understands the natural flows of life.

Instead, we are too easily distracted, indeed blinded by those bright flashing lights operated by the so-called movers and shakers, people who continue to exacerbate the world's problems while promising easy gold, repeating the mistakes of the past in their attempts to manipulate nature in order to expedite their selfish personal interests. Ironically, conventional perspectives of progress all too frequently, actually, embody a process of negative retrogression.

Let's progress by going back to the future ... generatively.

Let's understand ... and work with ... nature's simple lessons about change and life.

This is a book about change.
It makes no claim whatsoever to be a comprehensive work.
Indeed, I don't believe that it is even possible to write such a compendium.
Such a book would indeed contradict the very essence of this ubiquitous notion that enables life.

This is not a book about change.
It's a book about life, which is actually the same thing.
The book makes no claim whatsoever to be a comprehensive work.
How could anyone be so ostentatious as to even think they were able to write such a compendium?

I am certainly not claiming to understand life.
I understand what I understand about life, which does not mean I understand life.
What I understand about life is continuously changing … as I learn.
I do, however, consider that I have acquired some understanding of change and the role it plays in our lives.

There are no contradictions whatsoever in the above sentences.
Paradoxes perhaps; but not contradictions.
Life is full of paradoxes.
Can we play with them?

I would love to think that people working in organisations and societies everywhere, regardless of size, feel able to use the principles and frameworks presented here to shape thriving healthy communities in which they live and work.
This challenge ought not to be all that difficult, yet it is clearly beyond the grasp of many organisation leaders. This is an observation, not a criticism.
We can change this … together.

For centuries, we humans have proven proficient in making life difficult for ourselves.
Let's change this too.

Throughout this book, I use the metaphors of nature, music and dancing to illustrate the notions of change and healthy communities.

Unfortunately, moving and thinking in accordance with nature are skills that appear to have become dormant among the majority of human beings. We were actually born with these skills. After all, we are nature. In the words of ecologist Rex Weyler, *"We are wild animals."*

Our social systems, together with their associated political and education systems, teach us to perceive our world as a machine to be manipulated. Therefore, even when we are designing initiatives with a generative intention, we tend to build in hinderances and stumbling stones that prevent us from dancing to nature's tune.

We need to relearn nature's flows; not merely learn about them academically, but internalise the principles so that working in accordance with them becomes natural for us.

Current estimates place the global population at 7.5 billion human beings. Every single human being gathers a unique collection of experiences during their life; experiences which are experienced through the lenses of those already acquired.
And so, our perceptions and perspectives of life are changing as we learn.
What a rich and powerful source of learning, evolution and development this is.
Let's cultivate it.

Throughout this book, I am inviting you to reflect on your own personal perceptions, perspectives and experience.

The text is interspersed with pauses, interludes and Questions.
These too constitute an invitation for reflection.
You, of course, will decide for yourself whether or not to actually make the time and space to answer the Questions.
Having learned that confronting uncomfortable Questions constitutes a rich source of personal learning, I would recommend doing so, while actually writing down whatever comes into your mind. The act of writing on paper, rather than a piece of electronic equipment, can help

to clarify what may well be a collection of confusing murky thoughts. Should your thoughts remain murky even after writing them down, that's fine. In the words of Alan Watts, the philosopher,

"Muddy water is best cleared by leaving it alone."

Outline, structure and content

I feel that the structure of a book about organisational and community change should reflect the theme.

However, nothing whatsoever about change is clear and structured.
Yet, a book of this nature requires clear configuration.
Decades of working with both individuals and groups has taught me the necessity of providing orientation within a supporting framework. When the context is open to so many different interpretations, as it is here, providing such a structure becomes even more important.
We have a dilemma.

Change is messy, occurring in interactions that cannot be readily defined, delineated or necessarily specified. I am trying to reflect this "undefinability" while doing justice to the message, by weaving through multiple contexts simultaneously and continuously.
Hopefully, the result remains readable.

The first section plays with the nature and ***essence of change,*** while dismantling many conventional misleading and damaging narratives that pervade and influence our lives.

We then take a brief excursion to familiarise ourselves with
Nature's Dance.

Dance Stoppers challenges a number of the popular myths, models and illusions that we believe constitute reality.

We do create our own reality.
And herein lies the problem.
The problem is not the fact that we create our own reality.
This is perfectly normal.
The problem lies in our belief that these myths, models and illusions actually represent the ways of the world.
Failure to recognise our beliefs as ... mere "beliefs" ... constitutes a self-constructed stumbling block, ultimately leading to poor decisions and failure of change initiatives with positive intentions.

The negative consequences arising out of the widespread belief in the validity of those ***Popular and Populist Myths about Change*** that pervade organisation life and society at large, render them too important to remain unchallenged. Here, they are afforded their own extensive cross-examination.

We then apply our insights to the essential Question that ought to concern organisation leaders everywhere:
How can we Cultivate Thriving Healthy Organisations?

We dance through the fields of culture, health, business, change, perceptions, perspectives, consulting, communication, organisation change, organisation development, organisation design, ethics, politics, relationships, leadership, conversations, human needs.

These "fields" are, in fact, one field.
There is no separation.

Having clarified the holistic nature of Organisation Health, we use this as the foundation of a framework for Cultivating Healthy Organisations, a framework which any organisation can, indeed needs to adapt to its own unique context.

This is not a "how to" book. There are no pre-designed formulae or models. Such a book would contradict nature's principles. Frameworks provide orientation, available for each entity to use in creating its own unique "how to".

My experiences have taught me that this is by far the most generative approach.

Transitioning to Action provides further reflection and Questions for consideration prior to undertaking any new initiative.

Style

Essentially, you will find here a collection of challenges, reflections and Questions.
As such, the sections are brief.

Both grammar and style pay no heed to convention.
Not all dances flow smoothly.
Sometimes the music changes in the middle of the party.

What do you understand by Change?

Change may not be what you think it is …

… but then again, it may well be, as I have no way of knowing who you, the reader, are at this moment in time …

So, before continuing to read, I invite you to reflect on this Question …

Qs *What do you understand by "change"?*

Change is possibly the most broadly misunderstood concept in society at large, and particularly in organisational life.

Misunderstandings about the nature of change frequently lead to corporate and organisation initiatives that not only fail to achieve their intentions, they may well actually cause more harm than good. If you are a leader in any kind of organisation, I imagine this is something you would like to change.

You can.

For each of us as human beings, for each organisation, for each community, change is natural and integral to life.

Change is ubiquitous,
it is quite simply a natural aspect of life.

Indeed …

All life requires Change.

Change is certainly not, as so often claimed, some"thing" that we "do" or even "have to do".
Above all, it is not "something" we "do" to other people.

In the following pages, I invite you to …

- … Question the conventional change narrative;
- … reflect upon and Question everything I say here;
- … reflect upon the true meaning of words frequently used when we refer to "change";
- … consider a natural approach to cultivating healthier organisations and communities.

To Question is fundamental to our learning and our development as human beings.

Questions live on, answers are transient.

'Question' is such an important word, it deserves a capital 'Q'.

If I had decided to write this book five years ago, the outcome would have been completely different.

Should I decide to revisit my writings in five years' time, the outcome may also be completely different.

Life changes us.
And let's be honest …
Who really understands life?
And should we even try?

In the words of Albert Camus,

"You will never live if you are looking for the meaning of life."

Feel free to join me in …

… Dancing with Change …

This is the only way I have been able to live my life … having learned though experience that it has no plans.

A Paradox

So, here I am, writing a book on change.

And at every moment …

… my thoughts are changing …

… which means the book is changing …

… regardless of whether I have committed these new thoughts to paper …

When I hike in the mountains, a new vista opens before my eyes every time I round a bend.

Confusion about ... Change ... and the Essence of Life

"Nature is trying very hard to make us succeed,
but nature does not depend on us.
We are not the only experiment."
(R. Buckminster Fuller)

Change is one of the most fundamental life-giving phenomena of our world.
It may also be one of the most widely misunderstood.

We urgently need to develop a more accurate understanding of change, one that reflects its true nature. Only then will we be in a position to cultivate a world that thrives, rather than the one which we are currently shaping, in which exploitation, manipulation and conflict have become the defining features.

We urgently need to develop a more appropriate model of leadership, one that reflects the collective responsibility we have for cultivating the environment in which we live.

We have learned to view the world as a machine containing a collection of parts (resources) which can be tweaked and adjusted (changed) to produce outcomes (goals) that we have previously defined as desirable. We design the vast majority of our institutions and organisations accordingly.

Nature works differently.

How did we choose the wrong metaphor for life's dance?

Two decades into the 21st century, only a fool would deny the scale and complexity of problems in our world. Wherever we look, we see so many countries governed and/or significantly influenced by pathological dictators, many of them "elected" by their people. Democracy can be (mis-)interpreted in so many different ways. We are destroying our environment, despite having access to better know-how, science, technology and education than ever before. Poverty is widespread, despite having access to better know-how, science, technology and education than ever before. On every continent, millions of refugees, each single one a human BEing, are fleeing war, oppression and famine. Psychological stress abounds in working life. This list of shame is endless.

Many institutions around the world are making commendable and necessary attempts, both random and systematic, to alleviate the suffering caused by these problems. However, it makes little sense to examine and attempt to solve any one of the challenges by focussing on any singular issue in isolation. All are interrelated with one another.

Belief in and the wish for independent agency is both understandable and seductive.
In adhering to this belief of independence, we are falling into the trap of perceiving the world as a collection of independent parts.
It is not.

As an individual human being, I am inevitably a part of – not apart from – the environments and communities in which I live and work. Our world is whole. As such, it is relational.

Relationships entail interdependency.
The nature of relationships means they are continuously evolving.

Reflection …

Reflect on your relationship with your partner at the time you first met.
Reflect on your relationship with your partner now.
How is it different?

How are you different?
How is s/he different?

This all too common syndrome illustrates the interrelational nature of our world:

A corporation bribes a dictator, let's say in Africa or Asia or Eastern Europe, to permit them to (ab)use local labour to exploit the country's natural resources, while employing extraction methods that destroy the local environment to such an extent that the soil is no longer able to grow sufficient food for the community. People flee to other regions where they hope be able to eat, and live a life that deserves such a description.
Those people already living in regions where refugees arrive, attempt to protect 'their' area from the 'outsiders'. Conflicts may turn violent forcing even more people to flee further afield. Artificial borders are erected to 'protect' regions in which the population is able to live in relative comfort. Hungry refugees may become desperate enough for food to permit themselves to be exploited by unscrupulous businesspeople in their new temporary homes. Or perhaps they are too weak to defend themselves. The cycle continues.

Admittedly, this example is presented in rather simple terms.
It is, nevertheless, real enough, and hardly fits in a world that is thriving.

The nature of life defines that we ought to be thriving.
As a species, we possess the resources, wisdom, knowledge and technology to enable us all to live rich meaningful enjoyable lives.

"Who is responsible?" is a common Question when we find ourselves in a situation that is different from that which we desire.

Realistically, we are all responsible for the state of our world: we human beings, who for the past couple of hundred years have been actively creating political, economic and social systems that seek to manipulate so many aspects of nature, including our fellow humans. The roots of these systems go back much further.

Why are we not thriving? What's going wrong?

Fundamentally …
We have learned to perceive nature as a collection of resources ("things" to manipulate) to exploit for our own profit. We believe that the engineering paradigm describes how our world functions.

When we shape our societies, and thereby our institutions and corporations, on a model grounded in the ideas of separation, power over others, exploitation, quick fixes (which paradoxically do not, in fact, "fix"), it is barely surprising that the outcome is continuous conflict.

Challenges to the mechanical model of the world are by no means new. Still today, they remain a minority voice.

We can change this too.

Origins of the machine metaphor to describe our world

The work of philosopher Rene Descartes (1596-1650) and scientist Isaac Newton (1642-1726) was integral to the mechanical metaphor that has shaped western society since the 17th century, and, especially since the mid-20th century, also life in the rest of the world.

Modern business architectures began to emerge during the industrial revolution in the late 18th and early 19th century, with organisations structured as manageable, interconnected functional parts comprising the whole working unit.

Taylorism (scientific management), based on the work of Frederick Taylor (1856-1915), emerged during the late 19th century and could be described as an engineer's attempt to fine-tune the workings of his machine. Many of its principles are still prevalent in organisation life today.
The advent of digitalization, which encourages measurement of all elements of any process, has given Taylorism a renewed boost.

Digitalization, unfortunately, has seen the proliferation of that unrecognised, yet highly contagious virus: measurementitis. Measurementitis is the misguided belief that everything meaningful can be meaningfully measured, giving birth to a plethora of useless, indeed misleading, measurements.

"Not everything that counts can be counted,
and not everything that can be counted counts."
(William Bruce Cameron)

"Resources" and "Manipulation" are core principles in mechanical models.
Furthermore, they have always been key pillars of colonialism.

Any self-respecting organisation will today claim to put people first and declare their people as their biggest "asset". Note the fault line in the terminology.
Those organisations truly interested in aligning their activities according to these claims, would be advised to seriously reconsider their structure, as well as their operational approach.

Those that continue to believe that an engineered structure is appropriate for a business will inevitably slide into the trap of attempting to manipulate everything within their company as if it were a machine.

The mechanical metaphor is totally inappropriate for understanding and shaping our interactions in any society.

It underlies the widespread misunderstanding of the notion of change.

Living Systems

"I ask myself often:
Why is it that everything changes?
And only one answer comes back to me:
"That is how life is."
(Sogyal Rinpoche, The Tibetan Book of Living and Dying)

I am neither a biologist nor a physicist, and will therefore offer only a rudimentary overview of insights from a selection of renowned thinkers who recognise our world as a natural, living, continuously changing dynamic. Readers may explore the references further.

"What is life?", is a Question that biologists Humberto Maturana and Francisco Varela posed in their research. In response, they developed the concept of autopoiesis to refer to autonomous, self-organising, self-making systems. They state, *"the being and doing of an autopoieic unity are inseparable."*

Physicist Fritjof Capra refers to an autopoietic unit as the most elementary form of organisation: *"a system capable of sustaining itself due to a network of reactions which continuously regenerates the components – and this from a boundary of its own making."*

Physicist David Bohm refers to the problem of our perception of the world as a collection of fragmented compartments. In Wholeness and the Implicate Order, published in 1980, he says, *"the notion that all these fragments are separately existent is an illusion, and this illusion cannot do other than lead to endless conflict and confusion. Indeed, the attempt to live according to the notion that the fragments are really separate is, in essence, what has led to the growing series of extremely urgent crises that are confronting us today."*

In the same book, Bohm argues that we *"need to look upon the world as an undivided whole, in which all parts of the universe, including the observer and his instruments, merge and unite into one totality."* He speaks of, *"Undivided Wholeness in Flowing Movement"*.

Philosopher J. Krishnamurti, referring to life: *"Life is a movement in relationship."* He continues: *"Relationship is life, and this relationship is a constant movement, a constant change."*

Theologist and philosopher Alan Watts: *"The only way to make sense out of change is to plunge into it, move with it, and join the dance."*

Fritjof Capra on organisations: *"In every human organisation there is a tension between its designed structures, which embody relationships of power, and its emergent structures, which represent the organisation's aliveness and creativity."*

Russell Ackoff, a brilliant systems thinker, said, *"When science divides reality up into disciplinary parts and deals with them separately, it reveals a lack of understanding of reality as a whole, as a system."*

He also said, *"Systems thinking is holistic; it attempts to derive understanding of parts from the behaviour and properties of wholes, rather than derive the behaviour and properties of wholes from those of their parts."*

The corporation or institution or organisation in which you work is a living system, an organism.

Any system that incorporates nature is alive.
Humans are nature, therefore your organisation as a system is alive.

Each interaction between any two people within the organisation's system changes the dynamics of not only the relationship between the people involved, but simultaneously, the dynamics of each of their relationships with all other members.

We would do well to listen to and learn from the wisdom of those who understand.

I am suggesting that we could relatively easily begin to evolve the thriving societies and communities that most people probably want and certainly, deep down, need, IF we make the effort to begin to understand the true … natural … role that change plays in our lives.

We are automatically continuously shaping our lives, individually and collectively, consciously or otherwise.

It is a delusion to believe we can separate ourselves from our role in cultivating the naturally occurring changes in life.

I personally find the following example inspirational: a living example of people who are changing the narrative from separation, fragmentation and conflict to wholeness and peace. Today, they are working actively to support others cultivate more generative communities.

"In 2006, Israeli and Palestinian former combatants, people who had taken an active role in the conflict, laid down their weapons and established Combatants for Peace. The egalitarian, bi-national, grassroots organization was founded on the belief that the cycle of violence can only be broken when Israelis and Palestinians join forces. Committed to joint nonviolence since its foundation, CFP works to both transform and resolve the conflict by ending Israeli occupation and all forms of violence between the two sides and building a peaceful future for both peoples."
From the website of Combatants for Peace: https://cfpeace.org.

Throughout life, we are dancing, dancing to music which actually never stops.
Only the tune and the rhythm change … and these we can influence.

Dancing with Change entails Dancing with Life.

Changing the Narrative

Surely not yet another book about change?

"Do you really think we need another book about change?"
asked a colleague as I explained my idea.
Of course, we don't need 'another' book about change.

I do suggest, however, that we need to counter the widespread harmful hype and misunderstandings concerning the role of change in our lives: in the business environment and in the societies in which we live.

In the world of organisations ... business, institutions, not-for-profit entities ...

> ... propagators of this hype, who completely misrepresent the nature of change, generally as "Change Management", manipulate the social media to drum up popularity and business for themselves, while remaining oblivious to the problems they create in the communities in which they operate.

In the world of politics ...

> ... politicians and their sidekicks (aka influencers) manipulate the same media channels to create and strengthen their power base, doing even more harm to the societies they are supposedly representing.

In order to thrive, we need to understand the nature of change, how it pervades our lives and how we can organise our lives according to nature instead of continuing our failed attempts to manipulate it.

We need a new narrative of change, one that is aligned with nature's flows.

Understanding the nature of change is not rocket science.
It merely requires us to open our eyes … and observe … through different lenses.

Having outlined the concept for this book, I mentioned it to the Managing Director of a specialist academy for executive education in Germany. Enthusiastically, he immediately began explaining challenges he had heard from a number of their clients leading small and medium-sized enterprises. They were tired of the standard fodder of so-called change management that actually creates more problems than it claims to solve.

Organisation Change

The vast majority of my work is with organisations, predominantly companies and corporations, and this is the primary context in which the content, experiences and reflections of this book are placed: Organisation Change.

This context of organisation change necessarily incorporates personal change, because without it, organisation change would simply not be possible.

Organisation Change and Personal Change
are inseparable bedfellows.

To illustrate this essential connection, consider the following Question …

You hold a senior role in a company that has just appointed a new CEO.
In contrast to his predecessor, he favours an open approach to management, decision making and information flow.

1. *Which Questions will you need to ask yourself as a first stage to becoming comfortable with this new approach?*
2. *How will this new approach change your interactions with:*
   (a) your senior management colleagues;
   (b) those employees who formally report to you;
   (c) the company's customers?

Interlude
A Walk through the Forest as the Fog closes in ...

"99% of who you are is invisible and untouchable."
(R. Buckminster Fuller)

Can we see the forest through the trees?

Words are limited and limiting.
Our language is inadequate for describing Nature's Dance ... change ... comprehensively.

All learning constitutes change.

As a child, I learned to Question all assumptions.
So, I did.
It wasn't always comfortable.
I still do so.
It still isn't always comfortable.
Questioning assumptions is not merely necessary, doing so is essential. Without Questioning, meaningful intentional change initiatives are not possible.

When was the last time you listened to someone telling you something and thought, "This doesn't seem right"?

My approach to my work and client challenges reflects my nature. Therefore, whenever presented with seemingly easy answers to complex challenges, I naturally become suspicious.

Strong intuition is deep.
What is your gut saying to you?
Can you trust it?

Significant meaningful intentional change is always preceded by deep reflection and personal work.

Life is transition …

… change accompanies us throughout life …

We inevitably experience change during numerous transitions throughout our lives.

I seldom have a reason to mention the word "change" while speaking and reflecting upon my personal experiences in life.

Think of the last time a friend came to talk with you after separating from their long-time partner. How often did they mention the word "change" when talking about their challenges?

I have been engaged as a facilitator and/or coach in hundreds of organisation and personal change initiatives.
Rarely have I or participants mentioned the word "change" while working.
Never have I positioned the notion of "change" as the focus of work I do with groups and individuals.

Life is amazingly multi-facetted if we care to open our eyes, hearts and soul to its true nature.
This does not mean that it is always beautiful.
It cannot be.

Organisations are communities in which people work.

You must understand the whole of life, not just one little part of it.
That is why you must read, that is why you must look at the skies,
that is why you must sing and dance,
and write poems and suffer and understand, for all that is life.

(J. Krishnamurti)

Nature's Dance

Random musings on

...

...

the nature of change in life
with illustrations of everyday natural change

Growing
Life is Relational (1)

Qs

What is causing this tree to grow in precisely this way?

What is causing everything around this tree to grow in precisely this way?

… The interactions.
… The millions of organisms at play, continuously interacting and changing the nature of those bits we can easily observe …

… And then comes the sun and the wind and the rain, and the children who stop by and examine it closely, playing and shifting stuff, which inevitably changes the composition of the whole …

And …

Interacting
Life is Relational (2)

Every single person in our world has a unique collection of experiences with life …
experiences of interactions with other people,
with institutions, with nature, with places, with themselves …

Consider the following statement and how it applies to you personally …

There exists no part of my personal story that does not influence my interactions with others, and thereby, the group context and its environment.

My personal story is continuously evolving.

And so, we have ...

… Everyday Change occurring in Every Organisation …

… Everyday Change occurring in Every Community …

Qs *How often do you share your stories with people in your community, in your workplace?*

The Power of Personal Stories
Life is Relational (3)

Human beings have been sharing stories with one another for millennia. Indeed, this may well be the oldest form of conscious learning known to us.

When stories are personal, they can be an extremely powerful source for community learning, generating shared understanding.

When they are abducted from their original context and related with the intention of manipulating the listener, as unfortunately happens all too frequently, they become a toxic potion.

It is not merely important; it is indeed essential to use the power of stories discerningly.

While sharing personal stories, people are engaging in a mutual tour of discovery, learning, together, about one another's lives and how each sees the world.

All organisation interactions occur in relationship.
What a powerful platform personal stories provide for cultivating the truly generative relationships that encourage organisations to thrive.

Personal stories enhance the power of integrity – personal integrity and system integrity.

Consider an unsuccessful project on which you worked …

Qs

Before you started …
Did you share stories with your colleagues about your experiences, preferred ways of working, perspectives and understanding of your work … while listening deeply …?

Did you … together … actively consider the Question: "How are we going to cultivate an environment to ensure our project is successful?"

Qs

In this unsuccessful project on which you worked …

How might the relationships and environment in the project have been affected; and how might outcomes have been different, IF you had shared your stories and considered this Question?

Sharing personal stories opens up a unique relational playground.

Observing Nature Changing
Nature's Dance (1)

Lie on your back … observing the clouds in the sky.
See how the sky is never exactly the same from one moment to the next.

Sit next to a slow-moving stream.
Observe the flow.
The water flowing through any point in the stream is always different.
The stream is never the same from one moment to the next …
continuously changing.

Sit by the ocean on a stormy day. Watch the waves crashing onto the shore.
Observe the changes on the beach and in the sea ...
… the coastline continuously changing...

Sit by the ocean on a calm day. Watch the waves gently lapping the shore.
Observe the changes on the beach and in the sea ...
… the coastline continuously changing...

Sit on a hilltop on a windy day … watching …

Observe the pond as the fresh water flows in from the mountains
as the snows melt …

Growing and Evolving
Nature's Dance (2)

Think back to when a child in your family was a baby.
While s/he is the same human being,
s/he has grown, evolved.

I am the same human being that I was 20 years ago,
but … am I the same person?

You are the same human being that you were 20 years ago,
but … are you the same person?

Are you the same person that you were yesterday?

Flowing with Nature's Changes
Nature's Dance (3)

Since I woke up yesterday morning, I have had countless experiences, encounters, thoughts … and you have, too. When we meet later, we'll have a very different experience to the one we would have had, if we had decided to meet yesterday.

Every day is different. Such is life …

IF we were truly conscious of the essence of Change, we would surely shape our lives according to nature's laws.

Or … would we?

IF we really understood the nature of Organisation Change, we would work with this natural phenomenon to actively shape our institutions, organisations, corporations, companies and communities so that they truly come alive … for the benefit of everyone.

Or … would we?

There is nothing whatsoever about Change that we have to "manage". Even if we wanted to, we couldn't.

Embracing Life
means
Embracing Change

Are we prepared to stop trying to "manage" Change … and … Dance with Nature?

WHAT IF … we, as groups of humans, attempted to access what we truly "know"?

Perhaps … we would be able to recognise that we actually know more than enough about life, and how Change is integral within life … to be able to shape healthy societies, communities and organisations.

Unfortunately, we rarely permit ourselves the space to search deeply enough.

I suggest that this is something we can … and … need to Change.

Weaving and Learning

"I liked it very much because completely different levels and themes were addressed. It was a non-structured complexity of themes. And this is precisely how I feel in work processes." (An executive commenting on his experience with cultureQs® during a team integration initiative.)

The group highlighted how new insights in the group were becoming accessible through …

… the natural weaving of continuously changing frames in interaction, in relationship.

It is not at all easy weaving together the pieces of my constantly changing thoughts as I am writing this book.

It feels like the parameters within which I am framing my thoughts are changing all the time … which of course they are.

The multiple experiences of my life are … inevitably … intricately meshed with how I am experiencing my world at any particular moment.

Nora Bateson has coined the term Symmathesy, which she defines as: *"contextual mutual learning through interaction."*

… illustrating the dynamic of group evolution …

Learning with others entails change … naturally.

Natural learning …
… illustrates why change defies definition …

Experiences are continuously weaving themselves into our lives in an ever more undefinable kaleidoscope of ever less definable colours …

The End
Nature's Dance (4)

When life stops changing, we're dead.

… Let's stay alive … thriving …

With our thoughts we make the world.

(Buddha)

Dance Stoppers

The more humans delude themselves that they understand
the workings of the world better than nature,
the more likely they are to erect stumbling blocks
that hinder their attempts to improve it.

Dancing while stumbling is neither easy nor elegant.

Wandering Blindly into the Swamp

Why should anyone have a
problem saying, "I don't Know"?

We saunter through life ...
... accompanied by our assumptions ...
... our illusions ...
... allowing them to blind us to what really ...
... is ...

... and so ...
... these assumptions compost a swamp ...
... which we mistake for a fertile field ...
... into which we innocently wander ...

And even when we do find ourselves in a field that is fertile, in our attempts to manage what is not meant to be managed, we transform it into a swamp.

Whenever I drive from my home to the mountains and lakes, I invariably enjoy the scenery and colours given off by the variety of crops cultivated in the Bavarian countryside. And invariably, I find myself thinking about standard local agricultural practices, and asking, "Why would any intelligent person consider it a good idea to spray poison (aka chemical fertiliser) on the food that we eat?"
And the only conclusion I am able to reach is that we have different understandings of "intelligent".

Have you ever tried dancing in a swamp?

How does our belief in the truth of our illusions, all of them Dance Stoppers, prevent positive intentional change initiatives from being successful?

Which illusions?

- The illusion of Understanding
- The illusion of Knowing
- The delusion of Control
- The illusion that constructed Models actually explain the world
- The illusion that we need Heroes, usually in the form of Leaders, who are waiting to act as saviours in delivering us from evil
- The illusion that the Myths we believe about change actually represent the true nature of the world

Dance stoppers are effectively …

… interventions that inhibit or even prevent communities from thriving …

We can begin to source the origin of dance stoppers by looking at the inherent pitfalls ingrained in our use of language.

Languaging ...
How our Patterns cause us to, unintentionally, Stop Nature's Dance

Our patterns,
the patterns that make you you and me me,
solidly anchored in the swirling depths
of our experiences and our culture,
have become so deeply ingrained in our lives,
that we are rarely aware of the extent to which they influence
our thinking and behaviour.

Especially in our use of language, these patterns come alive.

Humberto Maturana and Francesco Varela refer to language as a verb: *"It is by languaging that the act of knowing, in the behavioural coordination that is language, brings forth a world."*

Are we able to observe how deep-seated patterns define the language that reflects our perspectives?
Are we able to observe how these perspectives themselves initiate how we use language, creating our world?

Patterns and language play with one another in the soil that is our social environment. As we attempt to cultivate this soil using unnatural fertilizers, the field gets bumpier as the moles scramble to feed on the decreasing quantities of natural nourishment.

Let's consider just a few illustrative examples:

The reasons underlying our habit of using ***mechanical metaphors*** to describe natural interactions were examined earlier. In the context of organisation change, we regularly encounter words such as build, construct, engineer, hard-wired, design.

Military narratives are prevalent throughout our interactions in society. We couch challenges in terms of "war", people are "troops" or "soldiers", operations are "the trenches", competitors are the "enemy" to be "fought" and "defeated", after which we become "victorious". Even environmentalists occasionally fall into the trap of using such toxic terminology.

Corporations as well as many institutions, even charities formulate a ***"mission".*** "Mission" has its origins in religious colonialism. This alone is sufficient to consider it toxic, even before we scrutinize its military associations.

The general connotation of ***"growth"*** is positive. After all, more and bigger can only be better can't it?
Well, this depends. Context matters.

For those unable to get enough nourishing food to eat, more healthy food is positive.

For those who examine their waistline as they get older, "more" is perhaps not so positive.

For decades, if not longer, most countries have worshipped the god of economic growth.
In which light do they consider the growth in exploitation, environmental pollution, greed, corruption, psychological stress … that frequently, but by no means always, accompanies this growth?

Context matters. Always.

Qs

When you refer to the growth of a company, what precisely do you mean?

Qs

Consider and note down circumstances in which corporate growth might be considered:
(a) Positive
(b) Negative.

In each case, reflect and note why.

The four examples above barely touch the surface of the vast terrain of language and its (ab)use.

What could we change by reflecting more intently on the language we employ to talk about interactions in our world?

Discernment is important.

Let's return to Maturana and Varela:

"Every human act takes place in languaging."

Illusions, Models, Heroes and Myths – same same ... but different

Illusions

"A man who is unconscious of himself acts in a blind, instinctive way and is in addition fooled by all the illusions that arise when he sees everything that he is not conscious of in himself coming to meet him from outside as projections upon his neighbour."
(Carl Jung)

We delude ourselves that we understand what we cannot understand.
We delude ourselves that we know what we cannot know.
We delude ourselves that we can control what we cannot control.

And so ... regardless of our positive intentions ... we fall into nature's traps, simply by ignoring its flows ... its relationships ... its wisdom.

Models

We are forever busy constructing models that purport to explain ...
... how the world works;
... how people behave, as defined and separated by categories that other people have constructed to supposedly describe them;
... how to design blueprints for our organisations.

Even as I write these words, I cannot but help think how absurd they sound.

The problem? Our belief that these models reflect the way the world works.
They don't.

The need for Hero Leaders

"A hero ventures forth from the world of common day
into a region of supernatural wonder; fabulous forces are there
encountered and a decisive victory is won; the hero comes back
from this mysterious adventure with the power
to bestow boons on his fellow man."
(Joseph Campbell, The Hero with a Thousand Faces)

We have created models of leadership grounded in the medieval concept of heroes leading their troops into battle to defeat all-comers. These models are anathema to nature's understanding of its world. They are unhealthy, ignoring the relational foundation of all community life.

Myths about Change

Myths infiltrate our world,
and all too frequently,
become such an integral part of our reality,
that we are no longer able to distinguish between the myth and nature.

In the work of organisation and society change,
the essential distinction between myth and reality has become
somewhat nebulous.

By acting as though the myths we have created about change are actually a true representation of natural phenomena, we do untold harm to our organisations, societies and above all, the people who work there.

Common Examples of Dance Stoppers in Action

I imagine many readers are familiar with scenes similar to these …

You're in a closed room, working with your project team to meet an unexpected challenge. Everyone is engaged. The atmosphere is relaxed, focused, highly cooperative, yet somewhat stressed due to the pressure. You are making progress.

The door opens and the project sponsor launches into an irate monologue on the need to overcome this challenge by the end of the day.

You're at a private party. It's early. People are relaxed, chatting, enjoying the music. A few are dancing.
Suddenly, the music stops. Somebody has decided they want to make an announcement. Nobody is really interested.

Your project is progressing well, in a cooperative, focused, fun environment. After 6 months, one member of the team leaves to take up an opportunity elsewhere. When her replacement arrives, she is presented with access to all information, accompanied by the simple instruction, "You're experienced enough. Simply pick things up where your predecessor left off."
There were no introductions; there was no formal handover; there was no integration.

Smooth progress metamorphosises into chaos, conflict and accusation.

… illusions of control, heroism, knowing …
… ignoring the relational nature of everything …

Dance Stoppers manifest themselves in the false paradigms and inappropriate narratives we have invented for our world. They appear as barriers, hurdles, myths and traps.

In Organisation Change Interventions, these false paradigms give rise to a somewhat absurd paradox:

With our intention of designing a worthwhile initiative, we frequently prevent it from being successful.

In our nervous desire to ensure success, we frequently ensure failure.

The Illusion of Understanding

So many people in the world are providing (wannabe) "solutions" for challenges they have never faced.

What is the basis for these answers, these "solutions"?
It is nothing more than an idea … an idea of what someone imagines the situation to be … founded on the individual's limited perspective of the world, which itself is founded on their limited perspective of life.

Such "solutions" are no more than illusions.

My experience of life is limited … as is yours … necessarily.
My perspectives are limited … as are yours … necessarily.

This is not a problem. It is the only way things can be. Such is life.

Experience is visceral.

While experiencing … anything at all … be it a conversation with colleagues, a problem at home, a scene in a market, whatever … we feel … understanding the experience through the entanglements of our personal lived experiences up to that moment.

Perhaps we also understand it through our mood and circumstances in that moment.
Mood and circumstances certainly influence our experience of an experience.

Perhaps we don't consciously understand the experience as we live it.

Personal understanding of a personal experience is necessarily limited, and may change with the benefits of additional experiences.

My interpretation of my experience does not describe the way things are. It describes the way things are for me … in a particular moment.

Intense personal experiences that are frequently "explained" by those who have never experienced such an experience include:

• Being a refugee;
• Being a divorced father/mother;
• Being in debt;
• Being poor;
• Living in (a specific country);
• Being an entrepreneur;
• Formally leading a group of employees;
• Leading an organisation change initiative.

There can be no true understanding without visceral experience.

Academic understanding, the cognitive understanding of a concept, cannot constitute true understanding of that which needs to be experienced.

We need to listen to … truly listen and hear … peoples' stories. Even then, we will only understand them through the limiting lenses of our understanding. We can never understand completely.

Even with positive intentions, our projects and change initiatives fail as we stumble over ourselves while inadvertently pretending that we understand what we cannot understand.

Repetition of the words "experience" and "understand" is deliberate.

A global conglomerate headquartered in Europe had won the bid to supply a new power plant in Asia. Intending to ensure a smooth start to the initial phase of the project, they sent an intercultural trainer to Asia to train local specialists in "how to do business with Europeans". The same trainer also coached the project manager on "how to do business" with citizens in that country.

However, she had neither project experience, nor experience of doing business in that particular country.
The outcomes of her training and coaching?
A project team who thought they understood what they could not understand, partly because the trainer/coach did not understand what she could not understand ... resulting in confusion among all about the communication and behaviour of others.
The project kick-off, always a unique one-off integration opportunity, was ineffective.

The dance never really started ...

Why did the dance never really start?

Firstly, intercultural training does little to help improve understanding among people from different backgrounds who need to work together. Groups need focused conversations with one another to shape the unique working culture appropriate to them in their specific context.

Secondly, the trainer was clearly unqualified for the task she took upon herself. Responsibility for her appointment and the ensuing outcomes, lies with the individual as well as with those who contracted her.

Thirdly, the project management worked with a fragmented perspective of their initiative – the project. The people were perceived merely as "resources". Management failed to pose appropriate Questions that would have enabled a holistic integrated approach to the project, encompassing the people, the technology and the local environment.

Fourthly, and most importantly … this fragmented approach did not enable people to form relationships. While we can function without relationships, we cannot work together generatively. We need relationship to generate understanding.

As so often in such cases, the cost of realigning the project, a cost which necessarily includes the wastage generated by the original inappropriate decisions, was far higher than an initial investment in appropriate measures would have been.

While a comprehensive understanding of project management is essential for managing complex projects, it is insufficient.

Senior Project and Programme Managers, those responsible for ensuring successful delivery of complex projects, require an understanding of the whole context of the project including all its interdependencies. Above all, they need a strong sense for the social and professional needs of team members, the project team members. Without this awareness, they are highly unlikely to initiate an integrated approach to their work.

While we cannot bring forth understanding without the personal experience, we can begin to form images through deeper enquiry …

We can only bring forth true understanding through personal experience.
None of us can possibly gain personal experience of everything we need to understand.
That's OK.
We don't need to.

We do however, need to learn to hear the intrinsic wisdom that others carry within them.
How? By sharing personal stories in a genuine dialogue space, in relationship.

This dialogue space is a deep listening space in which members of a group learn to expand their perspectives by sharing experiences, Questioning their own assumptions, engaging in community meaning-making … and so gain a sense for appropriate action.

How does this work?
We look more closely at the power of Dialogic Conversations in Part VI, Cultivating Healthy Organisations.

The Illusion of Knowing

"The greatest enemy of knowledge is not ignorance,
it is the illusion of knowledge."
(Daniel J. Boorstin)

"We know what we know, but we have no idea what we don't know."
I remember first hearing this seemingly logical missive while participating in an advanced facilitation programme back in the 1990s. It appealed to me at the time. I have since recognised the trap ingrained in such thinking.

Do I really "know" what I think I know?
I suggest not.
The illusion of knowing is an extension of the illusion of understanding.

If I think I understand what I cannot possibly truly understand, then I may well claim to know what I cannot possibly "know".

Let's first clarify the kind of "knowing" we are not talking about here.

This is not about knowledge of facts. Facts are banal, also frequently questionable, confused by the ease with which information can be manipulated.

This is not about scientific knowledge, which evolves as scientists expand upon their discoveries.

This is not about knowledge of the kind of hard numerical data commonly used as a foundation for decision-making in organisations.

Here, we are concerned with the syndrome of …

- "knowing" the answers and "solution" to any systemic problem or challenge, especially without having asked the deeper Questions related to the problem in its context;
- "knowing" what others "should" do;
- "knowing" what will be beneficial to others;
- "knowing" how a so-called "process" for human development will lead to particular outcomes. We'll return to misconceptions about processes later.

This syndrome is pathological.

Serious work with people, with organisations, with communities, requires the antithesis of "knowing".

Serious relational work asks Questions.

Unfortunately, there exists a widespread tendency among leaders and consultants to make far too many assumptions about human needs and behaviour. These assumptions … because they are assumptions … are merely illusions.

Many of these assumptions build upon dubiously conceived research presented as scientific; research that purports to explain behaviours of human beings, divided according to pre-defined constructed (assumed) categories.

When academic and institutional programmes present these assumptions as "knowledge", learners take them on board. These leaders and consultants are not to blame for believing what they have been taught.

Qs

Now ...

... IF ... leaders learned to ask generative Questions ...
... rather than believing in the illusion of their assumptions ...

... How would their organisation benefit?

What is the origin of this pathological need to appear to "know"?

The following, admittedly, gross oversimplification, may help us to understand how the exaggerated focus on "knowing" flows through the education system into organisations.

Successfully passing school and university tests and examinations requires people to regurgitate parcels of information and/or ideas in a manner that is pre-defined by the test overlords.
"Knowing" thus becomes associated with success.
Business Schools continue in a similar vein, passing on their "knowledge" to students, many of whom subsequently join one of the major management consultancies whose business model is founded on providing answers for their clients ... "knowing".
Corporations and institutions place their trust in these management consultancies to apply their knowledge in advising them appropriately.

Patterns run deep.
The dominant patterns with which we grow up remain ingrained, weaving the roots from which we take our orientation in this confusing world of ours. Consciously or otherwise, these patterns shape our perceptions, perspectives, attitudes, and behaviour.

The illusion, perhaps delusion is a more appropriate term, of "knowing" is a trap that prevents people dancing to nature's tune, and may even prevent them being successful.

Here is one example from which I personally benefitted:

I was surprised to win the assignment to consult the Board on the integration strategy for their latest international acquisition. My competitor had an excellent reputation and had previously been a coach to one of them. When I asked the HR Director, with whom I had a very good working relationship, why they had chosen me rather than the competitor, his response was enlightening. "She presented a proposal with her own pre-fabricated answers, whereas you outlined an approach in which they and their employees would find their own path. The rationale made perfect sense to them."

A reflection Question …

You are sure you have a good solution to the problem you are discussing?
The people who will benefit are not interested.

What might be the problem?
What could you do?

The Illusion of Control

"If you want real control, drop the illusion of control;
let life have you. It does anyway.
You're just telling yourself the story of how it doesn't"
(Byron Katie)

The illusion of control is an extension of the illusion of knowing, which itself is an extension of the illusion of understanding.

If I think I understand then I may well claim that I "know", so perhaps I will attempt to act as though I "know" in order to produce pre-defined outcomes. And if I act in such a manner, I may well want to control the process and people's behaviours to ensure that my preferred outcomes actually occur.

A cursory glance at the results of our attempts to control nature's natural flows should be sufficient to enable us to see the negative outcomes ... but only IF ... we choose to look and, above all ... choose to open our eyes and see.
For example, what are the consequences of straightening a winding river or clearing forests to plant crops?

They include, flooding, the destruction of natural flora and fauna, reduced soil quality affecting the quality of crops, and more ... all leading to unnecessary human, environmental and financial damage.

In addition to exploitation of the earth's natural "resources", humans' attempts to control nature are arguably the biggest single causes of the ecological destruction that we are wreaking on our planet.

This is not the path to the healthy environment that will enable us to thrive in life.

Planning is elementary for project success.
Control is essential for the smooth functioning of technology, as well as for efficient administrative processes.

However, when planning incorporates attempts to control peoples' behaviour, we are more likely to fail than succeed.

People are not machines.
When they work as designed, machines exhibit predictable behaviour.
Peoples' behaviour patterns, nature's behaviour patterns are far too complex to be predictable.

An additional factor that is far too rarely considered in the design of organisational processes and procedures …

Qs

How about the serious ethical issues entailed in the idea of controlling and designing behaviour?

Dictators simply love designing and specifying behaviour standards.
Managers, mis-guided by their belief that organisations function like machines, frequently fall into this trap.
Ethical health is a fundamental characteristic of a healthy organisation.

The following experience vividly illustrates the outcome of falling into the trap of cultural dominance, which frequently expresses itself through the wish to control peoples' behaviour.

The HR Director of a global consumer products retailer confidently strode into the room for the opening of the new high potentials programme.
He proclaimed: "During the next two years, you will not only learn a lot, we will also introduce you to the behaviours we expect and help you become leaders of the future."
"Really?" replied a young man from Turkey. "This is not why I agreed to come all this way to participate in this programme. I don't want to be told how to behave."

In this example …

Qs

Who stopped the dance?
The young man from Turkey?
The HR Director?

Consider the people in your organisation as well
as those in its wider system to be its soil.
The better you cultivate the soil,
the more fertile it becomes,
the more nourishment it provides.

A government-owned international development agency was facing problems in a number of its health and hygiene projects in Asia. Their challenge: the local population refused to involve themselves in activities.
"Which role do the village leaders play in the project management?" I asked.
"None," they answered. *"We are responsible for the funds and have to ensure everyone works according to our specifications. We provide the resources, then empower them."*
"These are proud people," I responded. *"You are guests in their village. If you don't invite their involvement in a project that is supposedly for their benefit, how can you expect them to engage with you?"*

Neither the programme manager nor project manager were willing to change their approach, so I declined the invitation to support them, seeing no possibility of making a positive contribution.

Some years later, I was again approached by the agency, this time by a different department. Both the challenge presented and the outcome were the same.

The paradox is enlightening …

By holding on too tightly to the levers of control, the organisation stopped the dance of the project, a project with positive intentions and obvious potential benefits. Ultimately, the initiative was a failure.

Had they been willing to let go of the delusion of control, they would have significantly increased the chances of a positive outcome.

The fundamental problem of the agency actually lies far deeper, embedded within its roots.
Its attitudes are steeped in colonialism. Models and approaches with such a foundation … never could … and … never will be able to shape healthy environments.

The Illusion that Constructed Models actually explain the World

"The map is not the territory."
(Alfred Korzybski)

It was early afternoon on a beautiful clear day, as we flew over the wide expanse of land to which we have given the names, the Yukon Territory and Alaska. As I looked down taking in the stunning views, I remember thinking, "There's a border down there. How absurd! Most of this border comprises a straight line drawn on a map. That's even more absurd!"

In our attempts to make sense of and represent the natural complexity of the world, we create abstractions in the form of models, rules, narratives, etc.

And then, we pretend that these abstractions explain the world, or at least a part of it.

With this pretence ... including the idea that the artificial divisions we create actually represent the true world ...
... we exacerbate the environmental, ethical, psychological, relational damage we have already foisted upon ourselves.

We seek clarity.
It is easy to understand the allure of clarity.
After all, daily life can be so confusing.
Navigating the multitude of contexts in which we live our lives today can be overwhelming.

Picture the challenges of organisation leaders …

… dealing simultaneously, with pressure from …
… owners, employees, society, business partners, customers, suppliers;

… while …
natural disasters, government regulations and political events outside their sphere of influence can create havoc with their strategy;

… while …
being held responsible for decisions that may pull at their core, the core which gives life to their ethics.

… Oh …
Did I forget to mention the needs of their family, as well as their needs for connection with their family?
And …

Survival can be a relentless challenge.

Any idea that appears to bring a semblance of sense into this whirlwind is naturally seductive. Especially on bad days, everybody has bad days, we salivate at the prospect of finally having an ordered framework for our lives, and perhaps naturally, grasp at models containing the promise of such order.

At some stage, however, we wake up from the dream in which we envisioned the illusion of the model that explains …
… as we realise that the model did not explain.

The map is not the territory.
It is not even an abstraction of the territory.
The map is an abstraction of an abstraction.

Nature is real, not an abstraction.
It contains neither straight lines nor neat packages.

Your organisation, a living organism,
functions according to nature's rules.
Nature is not a model.

How do the structures of our surroundings pervade our thoughts?

Models of government and governance everywhere occur within the framework of those entities we know as countries. Organisations are formally located in one or more countries, subject to its laws and regulations.

Yet there is nothing natural whatsoever about countries as entities. National borders are all, without exception, artificial constructs. Indeed, they change more frequently than we may like to realise.

What are national borders?
They are nothing more than the fragmentation of the world resulting from the power games, generally violent, that power-mongers play.

What is the origin of these power games?
Thought; violent thought.

How are organisations traditionally designed?
By dividing them up into sections which are considered "manageable" and treated as independent units. The division of the organisation mirrors the divisions that we have imposed upon our world.

This practice is riddled with traps.
The organisation remains an entire entity, a whole.

A note of clarification:
Work in organisations needs to be divided up … and managed.
The problems arise from the manner in which organisations tend to be divided up and the way work is managed; especially, from the way people in these organisations are managed.

Our problems originate in thought; the way we think;
the way we think about our world.
Speaking of the difficulty in thinking, David Bohm formulated this problem for us in his book, On Dialogue:

"One of these difficulties is fragmentation, which originates in thought – it is thought which divides everything up. […] In actuality, the whole world is shades merging into one. […] Later we give this separation great importance."

Nature is real, alive, not an abstraction.
It contains neither straight lines nor neat packages.
The same applies to every organisation.

Models for Organisation Life

Some years ago, I was staying with an old university friend during a business trip to London. He came into my room one evening, saw some documents that I was preparing for a meeting the following day, and commented, *"You consultants do like your neat triangular and rectangular models, don't you?"* I stopped … and recognised the trap I had stumbled into.

What do all models have in common?

The ideas underlying the models are nothing more than a mirror of the thoughts their architect … thoughts grounded in cartesian reasoning.

Most are innately reductionist, ignoring the dynamic interrelational complexity of the living organism that is the organisation.
Mechanical models to describe the workings of organisations constitute a misguided attempt to reduce this natural complexity.
It is impossible to reduce the complexity of the world.

Understandably, some models reflect approaches that consultants have successfully implemented with their clients.
To these consultants I would ask two Questions:

Qs

(a) "Who implemented what, how, why?"

(b) "Which criteria are you using to define success?"

Management consultants frequently superimpose their models upon the dynamic natural world, while declaring them to be universally applicable.

All models will find followers. All, regardless of positive intentions, will form stumbling blocks to the natural order of the organisms that are organisations, as well as communities.

I googled "Change Models" and received "About 8.220.000.000 results". Wow! There must be something good among these. Perhaps there is. However, I remain dubious.
The first 8 results in order of appearance were:
Major Approaches and Models of Change Management, 8 Critical Change Management Models, 7 Fundamental Change Management

Models, 10 Proven Change Management Models, Top Ten Change Management Models, 6 Essential Change Management Models, 5 Change Management Models, Models of Change Management.

Thought and its ensuing practice, together with algorithms (created by human beings), have converted "change" to "change management". The myth of "Change Management" will be negated in the next chapter.

What remains?

Thought and its accompanying Illusions …

… the Illusion of Understanding …
… the Illusion of Knowing …
… the Delusion of Control …
… the Misconception that constructed models reflect the natural world …

… in contradiction of nature's laws.

During a business trip to Japan, a professor at one of Tokyo's universities invited me to participate in a research project he was conducting into differences in communication styles between Japanese and western businesspeople. We were 12 people in total, and were initially asked to complete a questionnaire. The professor followed up by interviewing us individually, in the presence of the whole group.

He was becoming visibly more frustrated while interviewing me until, suddenly, he burst out, *"Eric, you're uncategorizable!"*
We all laughed.
When the group went for a beer together at the end of the day, I explained my background. The professor understood. He also recognised how he had become entrapped within the structure of the model he had devised.

So, what are these models that supposedly explain how organisations work or should work, and how people behave? And how are they commonly represented?

Let's look first at those that claim to be models of change processes (another myth about change).

Some are represented as a staircase; others as a linear flow from left to right; others as a collection of interconnected cogs.
A brief internet search will reveal further variations on these themes.

All are clearly presented, encouraging acceptance.
All indicate linear development and/or clear delineation.

Let's remind ourselves ...
Evolutionary processes are never linear.
Nature rarely if ever displays clear delineation.
Change is ubiquitous.

During the 20th century, a seemingly logical idea emerged ... that it is possible to create pre-defined categories that represent types of people.

I vividly remember a lunchtime conversation with my colleagues in New Delhi.
"There are two kinds of people in the world," began one.
"Correct," responded another interrupting him in full flow. *"There are those who think there are two kinds of people in the world. And then, there are the others."*
The laughter quickly spread to the neighbouring tables, then through the rest of the canteen.

Real and self-proclaimed psychologists have gifted us the wonderfully absurd idea that it is possible to meaningfully divide up people into categories according to sets of characteristics that they themselves have constructed.

These "characteristics" are commonly depicted as quadrants in rectangles or circles; perhaps as concentric circles; perhaps as fuzzy shades where the delineation boundaries meet.
All are merely variations on a theme.

The categories ... and subsequently the people to whom they are applied ... are assigned letters, colours, numbers, descriptive words.

And now ... bingo! We have a magically neat and easy way to describe people.

Except that we don't.
We have yet another illusion.

How about other Organisation Design Models?

In recent years, we have been able to observe the spread of models, representing ideals, developed on notions borrowed from political structures.

These models may be assigned titles ending in "-ocracy", a term that has its origins in the Greek word for power or rule.
In effect, supporters of such models suggest that their personal pre-fabricated notions of ideals, should be superimposed upon all organisations.

Even democracy has found favour among some consultants as a model for organisation design. While democracy, with its numerous imperfections, may well be the best political system we have invented to date, it necessarily incorporates winners and losers among the people, making it completely inappropriate as a model for healthy organisations.
Why ever would we want to have losers within our organisations?

Teal is a rarely used alternative English word for the colour turquoise. In the past few years, this four-letter word has become a popular descriptor for a movement that purports to have created the ideal organisation design framework, known as teal organisations.

A well-received book published just a few years ago, has contributed to the popularity of this trend. It claims, simultaneously, to present a new model for organisations, blueprints for the future of work, and that organisations who introduce these operate from a higher level of consciousness.

Personally, I have never experienced "consciousness" in association with models and blueprints.

I know I am not alone in observing the evolution of sect-like patterns in the behaviour of many supporters of this model. I wonder to which degree these people are aware of the pattern.

"Agile is an iterative approach to project management and software development that helps teams deliver value to their customers faster and with fewer headaches. Instead of betting everything on a "big bang" launch, an agile team delivers work in small, but consumable, increments. Requirements, plans, and results are evaluated continuously so teams have a natural mechanism for responding to change quickly." (Source: https://www.atlassian.com/agile).

In the context of software development, this process makes sense. Unfortunately, Agile is frequently presented as a model for developing organisations, for which it is completely inappropriate.
Living organisms do not evolve iteratively.
Software does not provide a meaningful representation of the world.

Today, in the first decades of the 21st century, technology is dominating business.
The social media has become a key influencer of people's perspectives

of the world.
Many Agile practitioners possess, or work with companies possessing, the technology skills required to manipulate the social media. Present day social media is, after all, designed to be manipulated. Manipulation of the social media contributes to the spread of the delusion that Agile is an appropriate organisation design model.

Broad-minded Agile coaches also recognise the dangers of such manipulation, while criticising the preaching undertaken by many of their colleagues, some of whom readily describe themselves as an "Agile Evangelist".
Evangelism is colonial.

I know I am not alone in observing the evolution of sect-like patterns in the behaviour of many supporters of this model. I wonder to which degree these people are aware of the pattern.

The distinction between "Agile" (iterative digital product development), and "agile" (flexible like a cat) is essential.

Without doubt, modern organisations need to be "agile".

Summarising our Insights about Models

The idea that models can describe our world is a mechanical notion. For technology, for technological and administration processes, models are very useful. For predicting outcomes of technological interventions, they may also be useful.

For organic interactions in organisms, organisations, they remain abstractions of abstractions and are more likely to result in confusion rather than clarity.

And now …

The only people who can create a generative model appropriate for their organisation are those within that particular organisation.

Any model evolved by the members of an organisation will be unique, applicable for their organisation alone.

Experiences shared by other organisations are always a source of rich learning, and we may well find ideas and approaches that are worthy of adaptation to alternative contexts.

Adaptation and copying are essentially different.

Orientation is essential, especially when undertaking any kind of change initiative.

Constructed Models merely provide the illusion of orientation, therefore, effectively disorientation.

Frameworks, on the other hand, are able to provide the orientation people need.

The chapter on Cultivating Healthy Organisations examines the characteristics that underlie organisation health, before presenting a framework for re-shaping organisations in line with these characteristics.

The Illusion that we need Heroic Leaders

I first worked with Daniel more than 20 years ago. Since then, we have cooperated on numerous initiatives in the various companies he has worked for. Now, as a member of the executive team of an established SME in the energy sector, he invited me to attend their annual open day for customers and share my impressions of Hans, the newly appointed CEO whose predecessor had recently retired.
I was shocked by Hans' condescending attitude as he interrupted presentations from senior experts to draw the audience's attention to himself. And, of course, he was going to lead them to unprecedented success through uncharted waters.
Within one year, the supervisory board had sent him packing. He had alienated far too many people.

We have already seen how the patterns of experience and culture embed themselves within us to weave new patterns that invariably influence our attitudes and behaviour.

In the vast majority of societies throughout the world, children grow up being taught to worship mythical beings in the form of an all-powerful god that supposedly guides, even determines life. Regardless of the descriptor we apply, be it God, Allah, Buddha, Brahma, Shiva or … (there are so many more), the syndrome remains the same.

In many, but by no means all of these societies, the doctrines attributed to these powerful mythical figures dominate everyday life.

Should we be surprised that people are perennially searching for heroic leadership?
I suggest not.
Patterns repeat themselves.

We are faced with a paradox here.
With our thoughts, we create our world.
So, our thoughts, more accurately, our ideas and beliefs about the world, create our reality.
As Paul Watzlawick's work has shown us, *"The most dangerous delusion of all is that there is only one reality."*

Beliefs are merely beliefs and do not therefore represent reality.
Beliefs create our reality.

Corporate history is littered with all conquering presidents and CEOs who have successfully led their organisation through unprecedented growth and won boardroom battles.
It is also overflowing with disasters, that those who appointed these all-conquering presidents and CEOs, quietly wish away.

Are we capable of learning from the disastrous consequences of believing in the myth of the heroic leader who will solve all our problems and lead us toward untold riches?

I remain hopeful, also confident, even while observing so many people ignoring history's lessons, as they repeat past mistakes.

I see colleagues describing themselves as Gurus.
I see coaches who "know" what their clients need.
I see consultants who have the answers to their clients' problems.

Heroes to a (wo)man.
Dance Stoppers to a (wo)man.

We all have so much to learn.

There is another way …

Henry, the newly appointed Managing Director, strode into the room in which all 127 employees of the traditional manufacturing company were nervously waiting.
"I'm going to tell it like it is," he said. "We're in the shit; deep shit.
"I have no idea who did what before I arrived and am not in the least interested in who may or may not have been at fault. The only way we are going to turn around this situation is together. We have to look at everything in the company and change what needs to be changed. I need your help. And I'm here to help you. Ask me any questions that you like, then we'll begin."

Henry kept his word. After 3 years, they were again profitable, having increased the workforce by 20%.

John Heider has produced a wonderful adaptation of the Tao, The Tao of Leadership, a book which I cannot recommend too highly. It is full of wisdom to reflect upon.

In chapter 10, Unbiased Leadership, he writes …

"Learn to lead in a nourishing manner.
Learn to lead without being possessive.
Learn to be helpful without taking the credit.
Learn to lead without coercion."

The Illusion that Myths about Change are Accurate Representations

The widespread belief in the truth of the many myths about change strongly influences the way in which we design the unhealthy organisations and communities that pepper our world.

They therefore deserve extensive examination.
In the next section, we investigate 12 of them in detail.

Before doing so, you may like to reflect on this thought from Albert Einstein, together with the accompanying Question …

Qs

"Reality is merely an illusion, albeit a very persistent one."
(Albert Einstein)

What is reality?

"Those who cannot change their minds cannot change anything."

(George Bernard Shaw)

Popular and Populist Myths about Change

We are experts in self-deception.

The omnipresent machine metaphor as a lens through which to describe organisational life is so far-reaching, that even people who reject it, frequently, subconsciously, fall into the trap of orientating their language and their work towards … solutions, fixes, causes, measurements, processes, management and more. Change, according to this model, is merely a "thing" to be done.

The outcome? …
A language of change that reflects mechanistic as opposed to organic thinking, with "Change Management" proudly heading the list. This language has given rise to a collection of popular myths about change, myths that have become populist, omnipresent in the social media as buzzwords or buzz-phrases highlighting the "wisdom" of those who "know".

These myths thus become an invisible burden, weighing down organisations everywhere because those who believe them are unable to distinguish between the myth as illusion, and natural interactions as they are evolving within their world.

Designing initiatives on the basis of the veracity of these myths cannot be a recipe for organisation health.
It is a recipe for further systemic confusion.
They comprise a perfect formula for designing fear-based manipulative organisations.

This is not the path we wish to travel.

Popular and Populist Myths about Change

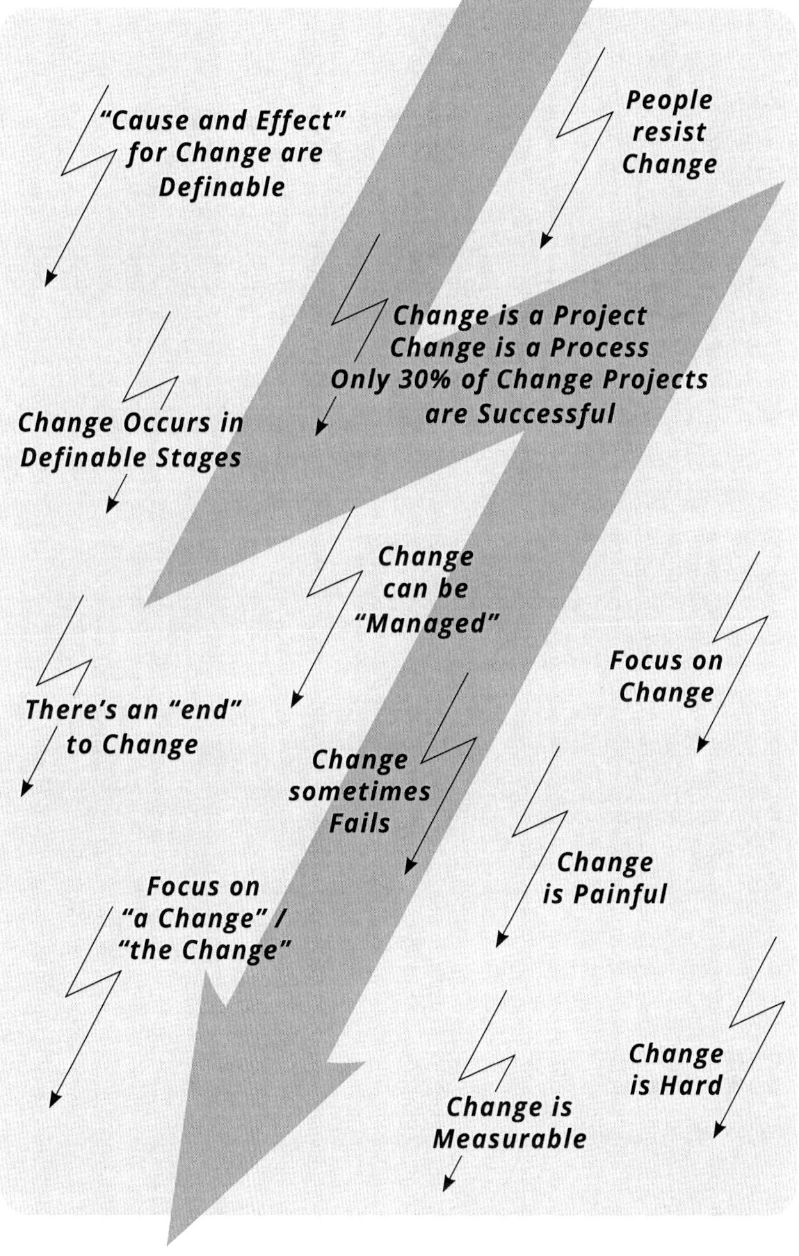

None of these myths are in any way new. All were alive and well during the 20th century.
Click-bait social media practice assists their dissemination.

They have gained such a strong foothold in society that, before demolishing and consigning them to the garbage bin, we might wish to remind ourselves of the true nature of change …

Change is ubiquitous.
It is the Essence of Life.
Dancing with Change entails Dancing with Life.

Let's examine each of these Myths about Change in detail …

1. Change can be Managed
2. Focus on Change
3. Cause and Effect for Change are Definable
4. People resist Change
5. Change is a Project / Change is a Process
6. Change Occurs in Definable Stages
7. Change is Hard
8. Change is Painful
9. Change Sometimes Fails
10. Focus on "a Change" / "the Change"
11. Change is Measurable
12. There's an "end" to Change

Myth ...
Change can be "managed"

The new CEO of a traditional engineering company was convinced of the need to encourage and promote a more open culture. We had just finished an intensive introductory session with 110 leaders and senior managers to kick-off the Culture Change Initiative. We used cultureQs (an accelerator that encourages people to engage in deep meaningful conversations very quickly), which brought to the surface not only the multiple layers of complex patterns in the room, but the realisation that every single conversation was changing the environment.
"How are we going to manage all this?", asked the General Manager of the Italian subsidiary as we were chatting during the coffee break.
"Wrong Question," I answered. *"Look at what you've just experienced. Do you really think trying to manage this initiative would be even possible, let alone successful?"*
"Ah. Now I see what you mean," he responded, smiling.

The notion of Change Management may well be the most widespread, misguided, indeed nonsensical myth about change to pervade the organisational world.

Merely the idea that we humans are able to "manage" the life-giving phenomenon that is change, is absurd.
The Cartesian world view still dominates our framing of society and organisation life, so the popularity of change management is perhaps not surprising, which makes it even more important to dismantle this harmful myth that damages so many well-intentioned initiatives.

Now, we do need to add an important qualifier here …
All projects need a process to manage changes in specifications and circumstances resulting from unforeseen problems and challenges – a functional process. This is change management, and in such a context, it fits perfectly.
Indeed, the idea of change management originated in project management.

The modern-day practice and hype of so-called "Change Management", combined with a plethora of certifications, is nothing more than functional project management principles applied to intentional change in the living organisms that are organisations.

In everyday working life, we manage work processes. Sometimes we decide to change them. In the new situation, we're still managing work processes, albeit different ones.
We are not managing "change".

An organisation may change its IT system. This is a project. Implementation of the new system may well lead to significant changes in working processes. It may also require people to learn new skills, generate new relationships, and more.
The successful transition from the old to the new system certainly requires management … project management. It does not entail managing "change".

An organisation may decide to change its structures, its design, its strategy, its priorities, its management, all of which will be visible to employees, and many to customers too. The decisions will result in administrative and operational changes, which will require

management – project management. Good project management will ensure that functions and operations continue to run smoothly under the new frameworks.
The "change" is not being managed.
It is not "the change" that needs to be implemented.

Successfully implementing a new organisation design certainly requires work … working together with people in the organisation to reshape it as a healthy entity.
This does not constitute "change management". It is Organisation Development.

We manage the affairs of our family and social life. Family conditions change: children grow up, partners may change, family members may move away, older family members die.
We continue to manage our affairs, not "change".

In all the above contexts, the idea of Change Management is an oxymoron.
We can define and manage aspects of the technical process involved in achieving a new desired status.
We cannot "manage" the human side of the initiative.
Humans are emotional beings, and emotions are unpredictable. In groups, the dynamic and chemistry of our interaction is even more unpredictable.
While we can certainly "manage" the manner in which we display our emotions, it is a fallacy to believe that emotions themselves can be "managed".

The quality of our engagement with other humans and our work reflects our emotions.

To summarise … While the concept of change management may make perfect sense for managing specification changes in projects and technical work flows, in the context of working with human beings it runs against the principles of living dynamic systems … and if you are reading this, you are very much alive and active in multiple interrelated living systems.

A reflection Question …

Note down 3 current initiatives in your organisation which involve a "Change Management" specialist.

What exactly are these people doing?
What exactly are they "managing"?

Myth ...
Focus on Change

I vividly remember a trouble-shooting workshop I facilitated some years ago for an international project team facing huge problems of their own making. In a nutshell: they were failing to meet targets; they had the wrong people in the wrong places; quality was unacceptable; the client was not merely unhappy, they were threatening to cancel the contract. They had to change and to make changes, and they knew it. My brief, negotiated with the head of the steering committee, was to enable the team to focus, quickly, on what they really needed to do to get back on track.
We were well aware that conflicts were inevitable.

The group's challenge and this assignment was all about making changes, significant changes. Yet interestingly, during the workshop, not one person mentioned the word "change". There was no reason to.

"Change" was not the focus of this intervention and quite rightly so. They focused on what they needed to do to deliver a quality product to the customer.

Change is not about "change".

A healthy change initiative is not about "change".

If "a change" is the primary motivating factor driving the initiative, I would strongly suggest reassessing the intentions beginning with the Question, "Why are we doing this?"
A healthy change initiative should be about the desired environment you wish to shape … in your specific context.

Any initiative that is merely about a "change", is meaningless.

Research over decades has highlighted that the most stressful experiences most humans are likely to encounter in their lives are, (a) the death of a close family member; (b) divorce. Both of these situations automatically bring significant change to a person's life.

Three decades ago, while going through a divorce, my whole life was in upheaval. The stability I'd imagined I had had, simply vanished. I was living change. However, I cannot recall even thinking about "change" at the time. Why would I? I was preoccupied attempting to make sense of my life, while bringing some order into it.

In 2004, I made a conscious decision to take a break from professional life, and leave Europe to live for a while in Asia. I sold or gave away every "thing" I owned. My decision was not about making a "change", but taking the rest I needed, while gaining new experiences.

When I was working closely with the Managing Director of a German engineering company on an initiative to reshape their corporate culture, we didn't focus on "change".
His appointment to the role, my visibility on site as an external consultant, the explicitly communicated intentions of our work, the manner in which we invited engagement from everybody ... all this constituted a significant change in the environment that employees and other managers were becoming accustomed to.

Our focus was clear and explicit, openly communicated throughout the company: "How could we achieve a turnaround so that the company once again became profitable, while evolving into a place where everybody – employees and management – enjoyed working?"

Generative change initiatives don't focus on "change".

A generative change initiative is about cultivating the desired environment which enables desired outcome(s) … without falling into the trap of manipulating the organism.

Here lies the focus.

Overcoming Problems – where should we focus?

Change ...	We don't focus on change to bring about change in society. It won't work.
Culture ...	We don't focus on culture to cultivate a healthy environment. It won't work.
Health ...	We don't focus on health to improve our health. It won't work.
Happiness ...	We don't focus on happiness to ensure a happy environment. It won't work.
Racism ...	We don't focus on racism to end racism. It won't work.
Profit ...	We don't focus on profit to ensure a successful business. It won't work (unless we're manipulators ... and that's not what we're interested in).

We focus on the whole entity ...
... Cultivating Healthy Communities ...
... so that the problems dissipate ...

A reflection Question …

You have been recommended as a coach to a friend of a friend who has recently, reluctantly, taken over as Managing Director of her family's 100-year-old business.
The company has a good reputation. However, demand for their products is significantly lower than in the past. They need to reassess their approach to the business.
She is unclear what she should do or where she should start.

Which Questions are you going to ask her?

Myth ...
"Cause and Effect" for Change are Definable

The Purchasing Director for the Consumer Goods Division of a global electronics corporation realised that better cooperation among colleagues located across five continents, could lead to significant savings as well as an improvement in product quality.

He engaged a colleague and myself to design and facilitate the initiative. Things began well. People thrived in connection with their colleagues from other parts of the world who they rarely if ever met. Indeed, as so often happens when human beings meet, they showed that they actually wanted to cooperate more closely.

The corporation's structure had evolved over a hundred years and could be described as solidly framed, if not rigid. Formal hierarchies were respected and most of those holding leadership positions expected the conventional outward signs of respect for their position.

Managers and employees developed a range of solid simple yet profound ideas for improving cooperation among the different departments. In order to fully integrate them, they needed the support of the director to help break down established structures. In addition, they needed his explicit authority to be able to take and implement concrete decisions. Outwardly, the director agreed ... and issued commands, which were subsequently ignored. He remained aloof, failing to follow up. His people lost trust and reverted to survival mode. The initiative failed.

We never doubted the good intentions of the director. However, in the end, he was unwilling, or felt unable to engage with his colleagues at senior level to confront the systemic issues in the corporation ... issues in which he was inextricably enmeshed.

The corporation's well-established crusty system was certainly a major contributing factor to the failure of this initiative. We cannot however claim it was the cause. Too many other factors were at play. The director himself: his attitude, fears, relationships and more, issues that were invisible to us as consultants. The power games that are inevitably played within traditionally structured organisations. Remuneration systems, including significant bonuses linked to "individual performance". In our internal review at the end of the initiative, my colleague and myself asked ourselves what we could/should have done differently at the start.

Overcoming technical challenges requires appropriate cause and effect analysis.

Overcoming human challenges entails incorporating the inherent complexity of living organisms into the approach.

Simple examples of technical challenges …
The car won't start; the lamp isn't working; the phone battery seems to run down too fast; the wi-fi isn't working; the air conditioning has failed; the neighbour's unfriendly dog comes through the hole in the garden fence.

More complex examples of technical challenges …
The automated production system is turning out inferior quality products; the corporate security system has been hacked.

For anybody with the appropriate skills and know-how, these problems with "things" are relatively easy to fix – find the cause(s), work out a solution, repair.
And in all these situations, such an approach is precisely what is needed.

While objects may be complex in design, they remain predictable.
Human beings are inherently unpredictable.

Consider the complexity in the following situation …

You were confused by your colleague's dismissive reaction to your suggestion in a meeting yesterday.

How was his reaction affected by ...?
... the conversation he had with his partner during breakfast;
... the interactions he experienced on the way to the meeting;
... his encounter with your brother at a conference last year;
... the Question his young daughter asked yesterday evening;
... the dispute he is having with the builders who made a hash of repairing his roof;
... the CEO's reaction when he presented the last quarter's sales figures?

Personal and Organisation Change Initiatives occur within systems that are alive.
No organism exists in isolation.
Interdependencies abound, and in their dynamic, are impossible to specify.
Everything is relevant.

Life's interactions are far too complex to reduce to concepts and processes that are designed for mechanical structures.

Organisation Change initiatives that perceive human beings as objects or machines (resources) "consisting of" a collection of components to be (re)calibrated, manipulated, repaired will fail … unless the only criteria of "success" are defined as either power or the financial result of the manipulation.

A reflection Question …

Qs

Take a walk in nature ...
Before you leave, tell yourself that you are going to observe as many details as possible of what you see.
These might be, for example: a crooked tree; new life emerging from a fallen, apparently dead tree; an entanglement of tree roots; flower buds where you may not expect to see them; plant life in water; etc.

Now ask yourself ... What causes these plants to grow like this?

Myth ...
People resist Change

In the early 1990s I was invited by one of the big three automobile manufacturers to join a group of eight trainers and consultants in Detroit. We were there to initially experience a new customer service orientation programme they had rolled out, and which we would then implement in German-speaking Europe. The corporation's intention was for us to take the programme as designed including all content. While the programme suited North American cultural patterns of the time, had we attempted to deliver it in the same manner in Europe, as the contractor initially demanded, it would never have been accepted. As a group, we rebelled and ultimately prevailed only when we gave the corporation an ultimatum: either you let us adapt it or we don't do it.

The atmosphere in the hotel bar that evening was tense. Someone suggested that the US management were resisting change. I suggested that they were resisting the need to give up their control of the programme they had designed as well as their idea that they "know" best.

"People resist change" is one of the most widespread and damaging myths about change that pervades society, and particularly, the organisational world. It is frequently employed as an argument to excuse people's failure to react to evolving situations that clearly demand a change of direction. Many consultants also use this fable as a means to present themselves as the heroes who overcome resistance.

The claim ignores the true dynamic nature of change, while circumscribing the real issues at hand.

Organisms are continuously adapting to new challenges and situations.
Such is life.
We learn.
New learning constitutes change.
When we stop changing, we die.
There is nothing to resist.

Let's look at two examples of situations for which this myth is frequently heard …

Internal Corporate Reorganisation

Employees may be unhappy with the situation and its outcomes and even openly express this dissatisfaction.
They will not resist the reorganisation – there is nothing to resist.
The decision has been taken.
They may resist being "forced" into a situation over which they have had no say.
They may resist the feeling of having to give up an element of control, itself a delusion, that they assumed they had over their working lives.
They may resist the consequences of the decision and/or the way it is being implemented.
In all these examples, they are not resisting "change".

Structural and organisational changes undoubtedly create challenges for employees.
While some people may initially feel uncomfortable with some new challenges that confront them, they adapt. Some may decide to leave.
They may deny the existence of the new reality, i.e. resist the fact of it.
They may even work against it, attempt to sabotage it. In such cases, it is the new conditions they are rebelling against, not "change".
Humans have always adapted.
They don't resist. Only adaptation to continuously changing circumstances enables us to continue living fruitful lives.

A serious problem requires action to be taken

Serious challenges, both personal and professional, may lead to a temporary paralysis on the part of those confronted with the problem.

Perhaps someone has lost a person they were very close to.
Perhaps a new technological development is threatening a company's established successful business.
And perhaps those affected simply don't want to believe this is happening to them. They are in denial.
They may be denying the importance of the problem. They may even be denying the existence of the problem … and initially do nothing.

In either case, they are resisting the truth of the new situation.
The environment of their life has changed.

Acceptance of the truth of the new situation as well as its significance, is the shift, the change that will lead to appropriate actions.

Denial of the existence of a new situation is not resistance to change.

People resist manipulation.
They resist when they feel forced by others to make changes
in a manner which goes against their natural instincts.
They resist push, pressure and above all, force.
They resist being forced to do something they don't believe in or for which they cannot see a good reason.

Examples of changes in community life include …

- The local market moves to a new location: some people may not like it, nevertheless, they continue to use it.
- The local transportation authority changes a bus route. Some people may not like it – some will be pleased – yet most will adapt and continue to use it.
- Your employer moves your office from one location in town to another, a few streets away. You may or may not like it, you will still be part of the move.
- A popular schoolteacher leaves the local school because he has moved to another city. Parents and children may not like the situation, but they will adapt.

There is nothing to resist in any of these examples.
People adapt.

Climate change and environmental destruction are among the most serious issues currently facing us as inhabitants of the earth. Overcoming these challenges requires huge integrated efforts across the world, including a significant change in attitude among all power-mongers, especially, politicians, institution leaders, industrialists, union officials, environmentalists, and others.

There is huge resistance to integrating activities, changing established practices, even accepting the scale of the problem. Obviously, misinformation campaigns play a part in the denial, but only a part.

I would suggest that it is not "change" that these people and institutions are resisting. This would constitute an easy cop-out argument for continuing to accept the current situation, including the ability to rant at those considered to be at fault.

Many perpetrators and activists are denying the interdependencies of the world. They are all protecting ingrained interests, working with ideas rooted in the pathologies of fragmentation and its ensuing conflict and greed.

They are not resisting change.

There is little doubt that organisations frequently encounter resistance to new initiatives.
What is this resistance? It is people resisting having to let go of the delusion of control over their accustomed working environment, together with all of its perceived predictable outcomes.

"Change" cannot be resisted. It is the oxygen of life.

The Kübler-Ross Five Stages of Grief Model – Beware of Misrepresentation

In 1969, Elisabeth Kübler-Ross first published her model of the five stages of grief to support those in grief while afflicted with terminal illness.

Unfortunately, this model is today frequently mis-represented as a so-called "change curve" and especially as an argument to support the idea that "people resist change".
Personal grief is a unique context.

This model is not a change curve.
It is a framework that reflects the author's learning on living, after the loss of a loved one.

Furthermore, people intending to present the grief curve in their work with organisations would do well to heed the words of David Kessler, co-author with Elisabeth Kübler-Ross of On Grief and Grieving:

"In our book, On Grief and Grieving we present the adapted stages in the much-needed area of grief. The stages have evolved since their introduction and have been very misunderstood over the past four decades. They were never meant to help tuck messy emotions into neat packages. They are responses to loss that many people have, but there is not a typical response to loss as there is no typical loss.

The five stages, denial, anger, bargaining, depression and acceptance are a part of the framework that makes up our learning to live with the one we lost. They are tools to help us frame and identify what we may be feeling. But they are not stops on some linear timeline in grief. Not everyone goes through all of them or in a prescribed order. Our hope is that with these stages comes the knowledge of grief's terrain, making us better equipped to cope with life and loss. At times, people in grief will often report more stages. Just remember your grief is as unique as you are."
(Source: https://grief.com/the-five-stages-of-grief/)

A reflection Question …

Think back to an event in your life that,
(a) made you feel uncomfortable … and …
(b) forced you to actively make changes.

- *What was the situation?*
- *What specifically made you feel uncomfortable?*
- *How did you react? Why did you react in this way?*
- *What did you then actually do?*

Expanding Perspectives …

Now … Ask one or two friends or colleagues to answer this Question for themselves.
Then, get together, exchange stories, listening carefully not just to what each person is saying, but also to how they are saying it.

If you are able to continue the conversation by diving more deeply into one another's feelings and reactions, you could, potentially, enter a space that is a rich source of learning about the patterns that shape your attitudes and behaviour.
Perhaps you will also experience this experience as learning … change.

Myths ...
Change is a Project. Change is a Process.
Only 30% of Change Projects are Successful.

For those who believe that change can be managed, it is natural to perceive change as either a project or a process.
We have already ascertained the absurdity of the notion of "change management" in the context of change in organisms. Let's remind ourselves ... organisms (organisations) are not pieces of equipment to be manipulated and therefore cannot be managed.

The consequence?
Change can neither be represented as a project nor as a process.

As for "Only 30% of Change Projects are Successful", this is nothing more than simplistic click-bait, a claim which we can regularly see in articles, books and on presentations slides.

Now ...
Let's get real ...

How could there be any such thing as a "change project".

A project has specific goals, a clear beginning and a defined end.
So how can "Change", which describes the essence of life, be a project?
Life has no specific goals. Life has no end, even though individual lives end of course.

Projects are set up with the intention of bringing something specific into the world, something that is new to the context. They have a beginning and an end, which makes them, by definition, temporary.
While the action of initiating a project undoubtedly intends to change a situation from that which existed previously, this alone does not

make it a "change project". The focus of a worthwhile project is the specific detail, not change in itself.

Digitalisation has opened up a plethora of opportunities to corporations, start-ups, digital product developers and others.
The opportunities presented by digital technology constitute a change in market conditions.
Digitalisation projects initiated by companies are designed to create new business models, introduce greater efficiency, obtain better decision-making data, etc.
Each project has a specific target which entails changes in the way the company operates.

Digitalisation projects are … digitalisation projects. They are not change projects.

Language, indeed all forms of communication, incorporate almost unlimited scope for misinterpretation, serious misunderstandings as well as misrepresentation.

Just as there is no such thing as a "change project", there is no such thing as a "change process".
A process follows a defined path through a series of stages until the endpoint / goal / transition point is reached.

A production, development or administrative process may be redesigned, necessitating additional changes in the way in which people work.
The process changes. Nevertheless, it remains a production process or a development process or an administrative process, etc. It is not a change process.

Healthy Organisation Change occurs through conscious "initiatives".

To be effective as well as healthy, Change Initiatives in Organisations need conscious and active cultivation.

Organisation Change involves continuous everyday work …
with the whole organisation.
It has no "end" … until the organisation dies.

How can we enable generative organisation change?

We take a deep dive into this essential Question in Part VI: Cultivating Healthy Organisations.

A reflection Question …

Qs

What is the one, most important thing, you would like to change in your organisation? (Be specific.)
Why?

Note down some initial concrete ideas that would enable your intention to actually come to life.

Qs

Assume your organisation accepts these ideas, and you begin to implement them.

What precisely are you initiating here?

Myth …
Change Occurs in Definable Stages

When plans get messy …

I received the call late on Thursday afternoon, 600 km from home, having just concluded a client workshop with 50 people.
The CEO understood that I would have no opportunity to prepare adequately for this initiative and that nothing more than a rudimentary briefing via phone would be possible. He nevertheless requested me to facilitate the launch as best I could. He had already outlined a plan for the process. Hmm … What could possibly go wrong?

The Board of a boutique international technology corporation where people generally enjoyed working, identified a new market opportunity. Wishing to move quickly, they founded a subsidiary. Luc, one member of the board was appointed CEO and assigned 40 highly qualified employees to transfer to the new company. All 40 employees were instructed to attend the launch workshop which intended to work on strategy implementation.

There had been little prior consultation with employees. Scheduling the workshop to begin on a Sunday evening was unhelpful. Luc outlined his strategy and had a clear time frame in mind for implementation. People had questions. Many of his responses were nebulous. He then attempted to brush aside an intervention from a young employee. The atmosphere became uncomfortable.

It was time to take a break.

We've already examined the illusion of models that supposedly outline steps to success for change processes, models that promise 3/5/6/7/8/… steps to the promised land. The books are popular, as are the teachings in business schools. Executives are willing customers, happy to bite on the bait … just as Luc did here, as he stumbled into the trap.

People are easily seduced by clarity … or rather … the appearance of clarity.
When we have an idea we wish to implement, we love to see a clear path outlining the process from the start to completion. The path, the plan provides a comforting sense of security that enables us to organise the upcoming weeks and months accordingly.

Now, for buying a house, replacing the roof, organising a conference, installing a new IT system, launching a new product, a plan including clear stages for implementation is essential. These are, after all, projects.

We have already ascertained that Change Initiatives, whether personal or organisational, are not projects; neither does it make sense to define them as processes.

Nevertheless, many leaders planning Organisation Change Initiatives often expect to see the outline of a pre-defined "process" prior to awarding consulting contracts. Many consultants readily oblige.
I too plead guilty to having fallen into this trap … before I learned the fallacy of such an approach.

Having worked independently since 1987, first as a business owner, and later as a freelance consultant, I certainly understand … the lure of an attractive contract, the need to make a living, the professional risk involved in asking uncomfortable Questions of the client. There may also be ethical dilemmas to confront.

Regardless, we would do well to remind ourselves that …

Life is a continuously changing web of interactions that,
in the particular moment they are occurring,
are themselves changing.

Now, suppose that a CEO or Division General Manager is intending to initiate a significant reorientation of the business, for whatever reason. Regardless of the details, in one way or another, the working lives (and possibly the personal lives) of all employees will be affected. Structures may change; technical processes and consequently, interactions and relationships will change; and more too.

While orientation is clearly required, we cannot realistically "plan" the details of the initiative.

People love clarity.
Life, however, is unpredictable!

The culture change initiative began very well. The CEO outlined her ideas clearly. The manner in which she participated reflected and emphasized the message she was transmitting. People liked and respected her. Notwithstanding the disruption that the new direction was causing, senior managers and line employees were actually enjoying working in the new open environment that was emerging. The third phase of the initiative had been outlined and we were already designing its introduction for multiple sites worldwide.

Then ... the company was acquired. The new owners were old school who valued control, and ordered the CEO to place the initiative on hold. Employees were frustrated and unhappy and sought meetings with the management team who were powerless to intervene. A number of well-qualified staff quickly found new jobs.

We can only design and work on Change Initiatives one phase at a time.

Once the focus and intention are clear, outlining a general direction is wise.

However, because every change changes the change, meaningful work requires an evolutionary rather than a planned process.

A reflection Question …

Think of a Change Initiative that you have been involved in.

What was the intention?
How did people claim it would work?
What actually happened?

Myth ...
Change is Hard

The relocation department of a global corporation had decades of experience in supporting employees and their families being sent abroad on assignment. Corporate employees using their services were, however, not happy. They felt their individual needs were not being taken into account sufficiently when preparing and planning for assignments, and were often left frustrated by routine non-committal answers they received to their questions, which only served to compound the stress of the experience for them. The HR Director asked me to help.

As part of a programme, I took all managers and employees, through a visceral experience that placed them in the situation of the person being sent overseas on assignment. They began to understand the needs of their clients. Consequently, they decided to change their approach and processes for supporting their colleagues.

A few weeks later, I held informal conversations with all members of the department. One of them said, *"Yes, it was initially hard getting used to having completely different conversations with our clients, but is has made our work so much more enjoyable and actually, easier. There's less stress all round."*

The challenge the people faced was not "change".
The challenge was getting accustomed to a new way of working and interacting.

"Change is hard" constitutes lazy thinking.
Human beings are naturally adaptable.
This is how we have survived on earth for thousands of years.

Experiences through which we may have encountered our own adaptability include ... children growing up and learning; young adults leaving home and discovering the beginnings of adulthood; becoming a parent; learning new skills in a job and working differently as a result; leaving one's home environment to start a new life somewhere else.

There will always be aspects of life that one person finds hard;
and another finds easy.

Adapting to a new situation may well be hard, which is completely different from claiming that "change" is hard.

A reflection Question ...

Qs

Think of the last time you changed your job or your profession.

What were the challenges?
What did you find easy?
What did you enjoy?

Myth ...
Change is Painful

This myth can be coupled with "Change is hard" as an example of lazy binary thinking.

The fall of the Berlin Wall in November 1989 reverberated around the world. Undoubtedly, it is one of the most significant global political events since the end of World War Two in 1945.
It brought joy and relief to (probably) the majority of the population in the GDR (East Germany).
It may well have been painful for the leaders of that country.
It brought joy and relief to (probably) the majority of the population in other countries that were previously satellites (colonies) of the Soviet Union.

The social and economic changes that followed were accompanied by pain for many people in the eastern states of Germany, as well as unimagined opportunity for others.

The ensuing pain to a significant portion of the East German population resulted from the manner in which West German politicians steamrollered their political and economic system and interests onto the eastern states.

The political changes for East Germans were accompanied by widespread abuses of public funds, a lack of interest in the welfare of the population, in an economy that was effectively bankrupt. Established economic and social structures crumbled.

"Change" was not painful.
Adapting to the new circumstances was painful.

In 1994, I was forced to restructure my business, taking a significant financial hit.
In the process, I needed to make decisions that were painful for me and for others.

The "Change" was not painful.
The situation was painful.

At the time, I had no way of imagining the new sense of freedom that I would subsequently feel as a result of confronting my business problems head on. I was equally unaware that this would later enable a range of new opportunities to open up.

A reflection Question ...

Qs

Think of two examples of major changes, both voluntary and forced, that you have made in your life.

To what extent were each of these changes ...
(a) painful? ... Why?
(b) enjoyable? ... Why?

Myth ...
Change Sometimes Fails

The myth of "change failure" is so naïve as to render it absurd.

Once we accept that change is the essence of life, the claim makes no sense whatsoever.

A change initiative may fail to achieve its desired outcomes.
The initiative may be considered a failure.
"Change" does not fail.

We are continuously transitioning through phases in life.
At times, we are conscious of these transitions, at others not.
At times, they feel positive, at others not.
Regardless of what happens, the transitions do not constitute "failure".

You may be dissatisfied with a particular aspect of your life, and decide to walk a new path.
Perhaps you then discover that this new path is not right for you.
Regardless of whether you describe this experience as learning or failure, "change" has not failed.

A business initiative my fail.
Business decisions may turn out to be misguided.
A change of job may transpire to be the wrong decision.
A new partnership may not work out.

Whatever the outcome, "change" has not failed.

A reflection Question …

Think of an idea you tried, unsuccessfully, to implement;
… or an unsuccessful initiative you were part of.

What exactly happened?
What exactly did not work?

Myth ...
Focus on "a Change" / "the Change"

Change is not a "thing".

Following their acquisition by a US competitor, the senior leaders of the established European biotechnology company met with the new US-American CEO for the first time. Leading up to this gathering, there had been a great deal of talk about "the change", during which employees and managers expressed concerns regarding the future.

We invited the group to participate in a visual timeline journey, highlighting significant events in their company throughout the previous 30 years. They then selected a few of the most significant and exchanged stories about the effects that these events had on the company.

They understood how their company had been changing throughout its life.
Talk about "the change" stopped. They began to focus on the new entity they wanted to shape together.

Change occurs in a context, in a relational context.

The only meaningful way to talk about change is within its specific context.

A reflection Question …

Is there anything significant you would like to change in your life now?
If so, what is it?

Why would you like to do this?
How do you anticipate that this will affect your life?

Myth …
Change is Measurable

I do sometimes despair at the number of highly educated professionals all over the world that appear to be afflicted with the little-known, but widespread virus of measurementitis.

What is Measurementitis?
The misguided belief that everything in the world can be meaningfully measured.

Are there any known antidotes to measurementitis?

A scientifically-proven cure has yet to be published.
However, the following experiences may result in a change of perspective that could render the virus inactive …

- A walk in the forest while it is raining.
  Notice how the air, the plants, the animals are changing.
- A conversation with your partner about love.
- Observing children as they play and experiment and learn without adult guidance.
- Observe how you feel after a swim in a cool lake on a hot day.
- Observe your energy level having just walked to the top of a mountain on a beautiful clear day.
- Notice how you feel while tasting a delicate sauce made by an expert chef.
- Observe your friends' smiles as you greet one another.
- Observe your colleagues' smiles in the morning as they greet others with whom they enjoy working.

… all things that truly matter in life …

Here we experience change in abundance.
And nothing whatsoever that can be meaningfully measured.

While writing this piece, a storm has just forced me to leave the terrace and continue working in my office.
It feels completely different, and yes, it is affecting my focus.

The Executive Team responsible for a technology business with an annual turnover of $2 billion were stuck. There was no obvious problem. They got along with one another well enough. All were well-qualified for their roles, all were open-minded. The business was healthy. Yet there was "something in the air" during the strategy workshop, something that nobody could pinpoint.

We gathered in a dialogue circle.
I posed the Question "What's going on?", and remained silent.
And they began to talk. Slowly. Gradually, the atmosphere changed. The tension dissipated and the conversation flowed more freely. Laughter returned. Focus returned.

Nobody was able to identify the actual problem. It lay in the liminal space surrounding ... the Executive Team at work, their relationships, the business, the multiple undefinable contexts in which all had arrived at the workshop, the place ... and more.

There is no meaningful measure whatsoever to meaningfully describe the changes that occurred here.

It is essential to counteract the narrow-minded harmful attitude that claims if you can't measure something, it is not worth doing.

How?
By consciously considering what really matters in life, including the community life that is organisation life.

A good start would be for societies to seriously revisit, reconsider and change the manner in which they educate children at schools, and adults in universities and business schools.

Learning is relational.
Change is a relational phenomenon.
What really matters is not about numbers.

It is possible to create a measure for anything at all. We frequently do. This does not necessarily mean that the measurement makes sense.

In a relational living environment,
whatever one might be measuring, it is not "change".

It is possible, indeed advisable, to compare business results prior to the start of a Change Initiative with those later. The results do not measure change.
While they do provide some indication of the impact of the initiative, even here, the numbers provide merely … an indication, not a comprehensive measure. So many additional external factors will

affect the results, making a direct cause-effect comparison impossible. Businesses need measurements for a variety of reasons.

I would recommend that, before introducing them, companies exercise greater discernment, asking themselves these Questions …

- *What precisely are we measuring?*
- *How are we measuring it?*
- *Why does it make sense to measure this?*
- *How will the act of measuring benefit the business?*

The opposites of these Questions are equally relevant.

A reflection Question …

If you are an employee, write down the KPIs on which your performance at work is being measured.

For each one, consider …
What precisely is really being measured?

Myth ...
There's an "end" to Change

Many Organisation Development consultants will be familiar with the following client conversations about planned Organisation Change Initiatives ...

Client: "So, how long will it take to complete this initiative?"
Consultant: "Experience tells us that two years is realistic."

Client: "So, how long will it take to complete this initiative?"
Consultant: "We should be able to manage this in 6 months."

Both conversations perceive change initiatives as projects.

The first answer may well come from reputable consultants who have fallen into the classical trap of applying "Change Management" project-based principles to organisation change.

The second answer is a common response from fly-by-night consulting companies claiming expertise they do not have. Every profession has its charlatans.
One can even find claims from "consultants" claiming to be able to do "organisation transformation" in 3 months.

Developing Antibodies to Myths about Change

How can we develop antibodies to these myths?
By giving ourselves space for reflection ...

Change …

"No man ever steps in the same river twice, for it's not the same river and he's not the same man."
(Heraclitus)

Change …

… With every breath, we breathe new air ...

… Blood flows through our bodies continuously ...

… As human beings, we are evolving continuously ...

… In the organism that is the organisation ...
… Every conversation changes the organisation …

… Change is the lifeblood of the organism …

…

…

…

… When Change ends … the organism is dead.

Organisation Change …

… A Generative Change Initiative has no end …
unless it is artificially stopped.

… Every organisation is continuously working to keep itself alive.

… Organisational life is what happens every day.

… A truly worthwhile meaningful initiative …
… entails …
… enabling new patterns to embed themselves in the organisation …
… patterns which …
… in every moment of every day …
… come to life …
…
…
…

… There is no "end" … until the organisation dies.

A reflection Question …

Reflect upon your life up to now.
What is the most significant change you have decided to make to date?

How has this decision enhanced your life?
How are you enabling yourself to continue to benefit from the outcomes of this decision?

"Dance, when you're broken open.
Dance, if you've torn the bandage off.
Dance in the middle of the fighting.
Dance in your blood.
Dance when you're perfectly free."

(Rumi)

Composing the Music

A Reflective Interlude

The nature of life defines that we ought to be thriving.

...

...

...

How can we Cultivate Organisations that thrive ...

... enabling us ...

... to dance to the tune of the natural life within?

Shifting Sands: Reflection (1)

"We confront the problem of understanding how our experience – the praxis of our living – is coupled to a surrounding world which appears filled with regularities that are at every instant the result of our biological and social histories.

The whole mechanism of generating ourselves as describers and observers tells us that our world, as the world which we bring forth in our coexistence with others, will always have precisely that mixture of regularity and mutability, that combination of solidity and shifting sand, so typical of human experience when we look at it up close.

The fact remains that we are continuously immersed in this network of interactions, the results of which depend on history."
(Humberto R. Maturana & Francisco J. Varela)

An Essential Question: Reflection (2)

What should be the purpose of an Organisation Change Initiative?

... to Cultivate a Healthy Organisation.

... or perhaps you have a different perspective?

Dancing: Reflection (3)

"God respects us when we work,
but loves us when we dance."
(Sufi proverb)

How can we dance at work?

We can choose the music we play,
which means, we can change it.

We can even compose the music ourselves ...

Towards Cultivation: Reflection (4)

Let's summarise …

Change is ubiquitous.
It's the essence of life.

…

…

Organisations are communities.
They're alive.

…

…

Attempts to manage or impose prefabricated models on organisation change initiatives go against the natural order and will only succeed where the intent is to manipulate resources for the benefit of the owners.

We're not interested in manipulation.

There is another way …

We can …

… Cultivate Healthy Organisations …

… that truly come alive and thrive.

Qs

What does the work of
Cultivating Healthy Communities entail?

How can we do it?

Let's see …

*"The greatest artist has no conception
which a single block of white marble
does not potentially contain within its mass,
but only a hand obedient to the mind
can penetrate this image."*

(Michelangelo)

Cultivating Healthy Organisations

Intentional Generative Change Initiatives ...

... driven by everyone in the Community ...

"Music is the language of the spirit.
It opens the secret of life bringing peace, abolishing strife."

(Kahlil Gibran)

Fundamental Questions

What do you want your Organisation to ... BE?

Moving Beyond the Limitations of What's "Possible"

It is possible for leaders to shape any kind of company to be profitable … and most do.
This is relatively easy.

It is possible for politicians and civil servants to shape any institution to work in a manner pre-defined by themselves, its masters … and most do.
This too is easy.

Profitability and functionality are important.
Functionality is always important.
For companies, profitability is important too.

Profitability and functionality will not, however, enable an organisation as an entity to thrive.
This narrow perspective may have provided an adequate framework for organisations in the feudal era, also in the early industrial age, periods defined by exploitation, during which powermongers generally displayed a disgustingly overt disregard for the human in the being.

For the 21st century, these criteria are insufficient, indeed taken alone, totally inappropriate, as a guideline for organisation health.

Leaders content to function within such a narrow perspective of organisation life are responsible for burdening their organisation with significant unnecessary costs: costs which are usually hidden, costs that invariably reduce the corporation's profitability, as well as service quality to their clients.

They are also responsible for the damage to the psychological health of many of their employees and consequently, their families.

No visionary, ever, was content with asking such a limiting Question as "What is possible?"

21st century societies need a different kind of organisation. And this requires leaders, those people whose position determines that they are formally responsible for cultivating their organisations, to ask Questions that move beyond the binary manipulative narrative that still today dominates corporate and institutional life.

By virtue of their position, all nominal leaders are the example for other people in the organisation as well as its wider ecosystem.

What kind of society could we have if these leaders asked themselves Questions such as …

What kind of world do we want to be living in?

How can our company / institution contribute positively to our society?

Everybody contributes.
Everybody is responsible for contributing generatively.

Let's see how we can enable this shift to occur …

Who am I without my health?

Qs

What do you understand by

... 'A Healthy Organisation'?

Re-Shaping an existing Organisation
to become a living healthy entity, means ...

Intentionally Changing the Organisation's Culture.

What does this actually mean ... in practice?

The Fundamental Question for any Organisation

Qs *"What do we want our Organisation to BE?"*

This is …

the most elementary Question
any leader or founder can ask themselves.

For a leader alone to ask him/herself this fundamental Question is essential, yet provides limited perspective.
The senior management also need to engage with it.
This too is insufficient.

Unless everybody in the organisation engages with the Question … together, answers have little meaning.

Health is a social phenomenon.
One could also make a strong case for involving key customers, suppliers and business partners too … as well as family members of employees in any initiative.

Why?
How else will an organisation thrive if it fails to take into account the needs of its people?
And ...

If an organisation is not consciously trying to thrive, why does it even exist?

Dear Reader,

Have you ever consciously considered this fundamental Question?

Have you ever invited ... your managers and employees ... the people in your community ... to reflect upon this Question?

Who else is going to answer this Question if not those whose lives are intricately intwined with it? ... The people in the Organisation.

Cultivating a Healthy Organisation ... is an inside job!

As I took on my first employees back in the late 1980s in what was to become my first company, neither this Question nor anything remotely similar, even occurred to me. I was simply trying to earn a living for my family and myself. I was well aware of what was important to me, which consciously and subconsciously, with my limited know-how at the time, governed almost everything I did.
Of course, many things didn't turn out as I'd intended.
It couldn't have happened any other way.

Seeing the Whole Organisation: Beyond Incomplete Frameworks for Organisation Health

"A human being is a part of the whole called by us universe;
a part limited in time and space.
He experiences himself, his thoughts, and his feelings
as something separate from the rest -
a kind of optical delusion of consciousness."
(Albert Einstein)

Western science has a tendency to seek "definitions"; perhaps a natural consequence of schooling in which teaching through fragmented thought patterns encourages us to divide the world into categories.

Not always are definitions appropriate for providing clarity.
Not always are they even possible.
Definitions incorporate an additional trap: they may well encourage people to think in terms of fixed entities.

Because organisations are alive, they are continuously learning.
Organisation Health is therefore an evolving field.
I suggest that "defining" the notion of Organisation Health would not be appropriate.

However, when working to (Re-)Shape an Organisation as a healthy entity, it is useful, indeed necessary, to provide orientation by outlining the crucial characteristics of a Healthy Organisation.

Organisation Health is a holistic phenomenon.
It must be: the organisation is a whole entity.

Personal health is also a holistic phenomenon.
Any localised illness or injury affects one's whole being.
Consider, for example …

The challenge of trying to think clearly
while suffering from toothache!

We notice a lack of health … viscerally.

Traditionally, organisation health has been measured in financial terms alone.
The idea of the triple bottom line (profits, people, planet) was introduced by John Elkington in 1994.
Additional models that have appeared in the intervening years include the criteria of process efficiency and customer satisfaction.

None of these frameworks provide a comprehensive picture of organisation health.

The following examples illustrate the limitations of conventional organisation health frameworks.

Armaments manufacturers constitute what is probably the most despicable industry created by humans. Weapons manufacturing corporations are highly profitable; customers are undoubtedly happy; it would not be difficult for them to apply offsetting accounting tricks to denote their "environmental friendliness"; those who choose to work in them are making a conscious choice; manufacturing processes certainly need to be high-tech and efficient. Yet ... corruption abounds, and above all, their products are designed solely to kill and destroy.

In no way can armaments manufacturers be considered examples of healthy organisations.

Today's large technology corporations are among the world's most profitable. They are undoubtedly innovative and some have implemented highly efficient processes to service their customers. Hundreds of millions of people use them daily, indicating a significant degree of customer satisfaction with their products and services, notwithstanding the millions who regularly complain about them. In their sanitised corporate proclamations, the importance of their people in contributing to their success is noted. Reports will be issued "proving" the positive effect of their policies on the environment. Many founders have also established "philanthropic" foundations. Both greenwashing and ethics-washing have become popular pastimes for corporate executives.
Yet for years, the press has regularly reported on what can readily be described as the toxic workplace cultures in many of these corporations. Choleric leaders, stress, mobbing, command and control practices abound. Wages and salaries range from luxurious to survival-standard.

These corporations cannot be considered as examples of healthy organisations.

A small private NGO focused on providing health care to disadvantaged communities in Asia and Africa had a well-deserved excellent reputation. Employees were engaged. Purpose and meaning were obvious. The communities they supported were grateful. Demand for their services was growing. Finding qualified employees was easy.
And then, the organisation folded. It's primary source of finance was a national government development agency. Political priorities changed; the financing dried up.
The organisation was dependent upon its financiers.

Dependency is not a healthy foundation upon which an organisation can thrive.

The first example epitomizes unethical corporations. Without ethical health, the organisation cannot be considered in any way healthy. In the second example above, in so many small medium and large corporations as well as public institutions, people are objectified; i.e. they are employed and expected to 'perform' according to pre-formulated guidelines issued by the organisation's decision makers. They are perceived and treated as mere resources.

Classical management literature refers to people as "Human Resources". The descriptor is toxic. Some literature goes even further, referring to people as "Human Capital" – an even more toxic notion.

In a large proportion, if not the majority of economically successful businesses, communications specialists are adept at sanitising environmental data to paint a beautiful green landscape, while proclaiming the critical contribution of their people to their success.

Sanitised data and marketing-speak do not render an unhealthy organisation healthy.

Which essential human principle is lacking in all of these examples, indeed in so many organisations everywhere?

... A principle that prevents such corporations and institutions from being considered as healthy, and prevents them from thriving ...

... Human Dignity ...

Human Dignity

What is an organisation without its people?

Is health possible without dignity? ... No!

Who is a human BEing without dignity?

What is life for human BEings without dignity?

How can an organisation thrive if people do not feel valued and accepted for who they are?

Human dignity is an ethical, psychological, social, environmental, political, and economic issue.

What does it take to treat people with dignity? Not much actually.

We return to the Fundamental Question ...

... *"What do we want our Organisation to BE?"*

... a Question similar to ...

... *"What kind of world do we want to be living in?"*

There is no essential difference between these Questions.

A Healthy Organisation has healthy ethics.
Its healthy ethics will be reflected in the psychological and physical health of employees, suppliers and customers.
They will also be expressed through its environmental decisions and activities … which will be reflected in engagement, product quality, and innovative capacity … all of which contribute to the economic health of the organisation, without which, it could not survive.

Quality of Interactions ...

The photo on the left shows a spot in my garden on a day in March. The photo on the right shows the same spot one month later. The changes are seasonal, and occurred naturally, without any gardening. Every day, various species of birds and squirrels come to play here. It rained; the sun shone; it froze over some nights; the soil is full of insects; the neighbours' vegetable garden, attracting its own variety of insects, was just behind this patch of ground. The environment is alive.
It is impossible to precisely specify which interactions were responsible for which visible changes; the interactions themselves constituted change, all occurring randomly and simultaneously.

The environment of the garden is continuously shaping everything within it while itself being shaped by all of its elements, themselves changing. It's healthy, as we can see.

This is how nature works.
This is also how organisations work.

A key challenge for organisations is ensuring that its interactions are healthy.

The Role of External Consultants: A Reminder for Organisation Leaders

Dear Organisation Leader,

"What do you want your Organisation to BE?"

Personal perception is shrouded in a confusing blinding mist.
You are inextricably enmeshed in the patterns of your own system.
This is natural. It could not be any other way.

If you are serious about inspiring the success of your organisation culture change initiative, you need the support of experienced external consultants, accomplished in the art of finer dialogic facilitation skills.
They are not enmeshed in your system,
They are able to view your challenges through a lens unaffected by personal interests,
They are able to evoke trust in groups, while creating and holding space for them.

However, ...

No consultant can provide a meaningful answer to the fundamental Question.

No dedicated "project team" can produce any meaningful work on this Question.

You cannot delegate your responsibility for the initiative.
And please, do not even dream of giving a manager a meaningless title such as "Chief Organisation Health Officer" or similar.
I am never completely sure from where such absurd job titles emerge.
I maintain the hope that their source lies in the relatively harmless effects of occasionally smoking something more far more potent than tobacco, or looking too deeply into their wine glass.

Cultivating Organisation Health is an inside job.
It requires the involvement of everyone with a stake
in the organisation's success ...
... Working Together.
The approach is relatively simple.
The intricacies are highly complex.

How could it be otherwise?
The issues and Questions are relational.

Everyone is responsible for actively shaping the space that will enable your organisation to come alive. You, dear Leader, are responsible for opening this space.

The Fine Print

To avoid any misunderstandings …
The following sections present a framework for an approach to Cultivating Healthy Organisations. A framework is not a model.

The framework presented here reflects the author's learnings from over 40 years' experience working in people and organisation development. This experience includes … learning from practices that resulted in unintended negative consequences, even disasters; as well as … learning from practices that have served him and his clients well. Naturally, it also includes his learnings in life.
The framework highlights characteristics and prerequisites for any organisation wishing to begin an initiative to re-shape itself as a Healthy Organisation.

Obviously, the nature of this framework is influenced by my personal ethics.
How could it be any other way?
Organisation Health is an inside job.
Human objectivity is a naive mythical construct.

Implementing the approach entails deep work.
It may well involve Questioning oneself and everything one has done in the past.
It may.
On the other hand, of course, it may not.

Academically learning the characteristics of Organisation Health is of limited use.
They need to be internalised.
Employees quite easily recognise leaders whose actions lack the integrity of their proclamations.
Cultivating a Healthy Organisation is an inside job.

To go fast … first … slow down.

There is no guarantee of success.
There never is, for anything.

Wherever you end up is the right place.
If it feels wrong …
… it's still the right place …
… indicating that it's time to reflect … and learn …
… how to change direction …

Cultivating a Healthy Organisation embodies intention to Change.
Intention to Change implies … Questioning … everything.
Nature is the guiding metaphor.
The systemic interdependencies are prevalent, forever alive,
themselves continuously changing.
Cultivation is care. The work is never finished.

We view the organisation as a whole entity … albeit not as an independent entity … rather as one embedded in a web, related to all other entities with which it interacts.

Who am I
without my health?

Characteristics of Healthy Organisations

Characteristics of a Healthy Organisation

"Don't play what's there. Play what's not there."
(Miles Davis)

Let's beware … Words are limited and limiting.
Words are merely words, all too often waved around thoughtlessly.
Furthermore, so many descriptive words, nouns, verbs and adjectives alike, mean different things to different people. They also evoke associations for each of us, especially those words that describe what we may refer to as "values". Context may also alter the meaning of a word.
Whether someone is speaking, listening or reading, it is the emotion behind the words that give them meaning.
When responding to particular words that other people use to express ideas, we would be wise to reflect on our gut reactions, clarifying the reasons behind the emotions they evoke.

Beyond words, there is the communication that neither wants nor needs them.
So much of significance remains unexpressed … verbally.

When working on Cultivating Healthy Organisations, we need to look far beyond the words into the spaces where interactions occur, visible and invisible interactions.
This liminal space is alive and buzzing.

In Cultivating the Organisation, we're composing the music.
So, as Miles Davis says in the opening word of his autobiography …
"Listen."

The Characteristics of a Healthy Organisation

Clouds play an essential role in life on earth. They help to protect us from the sun's heat during the day, while providing a blanket during the night to absorb the thermal radiation that the earth emits back towards space, preventing temperatures from getting too cold. Although we may sometimes curse the rain that the clouds release upon us, without it, we would not be able to survive. Clouds are essential to the health of our planet.

Organisations too need protective systems for survival, and what better protection could they grant themselves than to ensure they remain healthy.

In the diagram, the clouds depicting the key characteristics of organisation health also represent their relational qualities. They are always moving ... continuously changing both their form and their constitution. While a focus on improving any one single characteristic is important, it is essentially meaningless. Working to improve organisation health entails working on the organism as a whole entity.

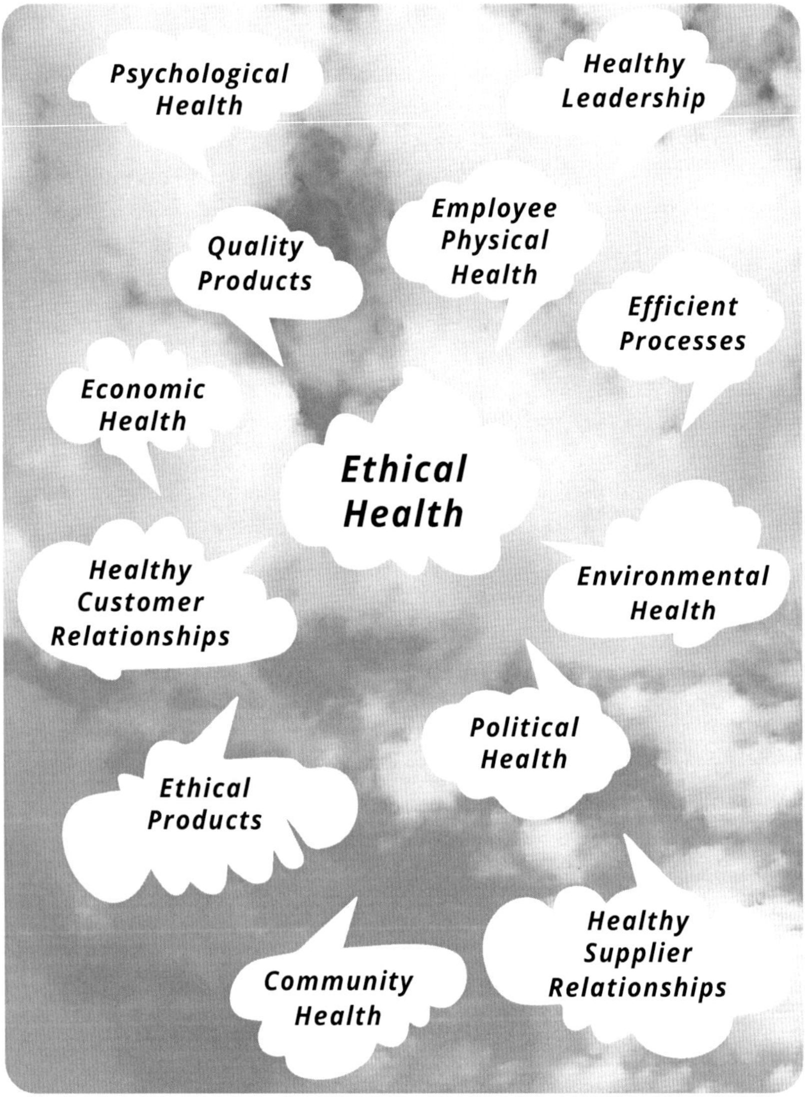

Let's look at each of these characteristics of a Healthy Organisation, and how they are intricately intertwined.
We'll begin in the middle with ethics.

Ethical Health

What do we mean when we talk about ethics?
Ethics are the internal reference points, based on values, that provide orientation for individuals and communities. Of course, everyone has ethics, even though, at times, these may be unrecognisable as such to those of us who value life, humanity, the BEing in the human, connection, relationship, dignity.
The subtle distinction between ethics (internal) and morals (proclamations to the outside world) is important.

Fundamental ethical Questions for every organisation include … What does the organisation stand for? How does it perceive its employees? How does it treat employees, suppliers, customers, partners, the local community, family members of employees? How does it interact with politicians and authorities? How does it source products? How does it make standard business decisions? How does it make difficult decisions? What kind of working environment is the organisation cultivating? … and above all … the fundamental Question … "What do we want our organisation to BE?"

Answers to the above Questions will also provide indicators to these Questions …
Why do employees choose to work there?
What do they say to themselves on Monday morning?

Let's consider just two Questions that highlight the interdependencies between ethical health and other characteristics.

Qs

How do the organisation's business decisions effect its employees, and influence the degree to which they decide to engage with their work?

How do decision-makers react to employees who voice ethical objections to working on specific projects?

The founder and sole owner of a successful software development company enjoys talking about the "ideal" working environment he has created. In addition to nonsensical proclamations that his company is free of hierarchy, he proudly announces that employees are free to select or reject projects on which they work.
"What happens when an employee rejects a series of projects for ethical reasons?", I asked during the workshop.
"If they don't want to work with us, they obviously have to leave," he replied.

While it may not be for an outsider to moralise upon his ethics and standards – it is after all, his company – we may still ask … "Which criteria drive his decisions?"

Economic Health

An economically healthy organisation is able to finance itself, without relying on subsidies or handouts in any form. Clearly, its processes need to be efficient, meaning it is not burdening itself with unnecessary costs. Customers want the company's products and are happy to pay the prices they demand. Employees are working effectively, assuming responsibility, enjoying being part of the community and engaging to ensure that they are adding value for all. The organisation and its suppliers work in partnership, ensuring that all are content with the value and quality available.

Government institutions including public education institutions, huge and vitally important employers throughout the world, clearly cannot be self-financing. Here, we might adjust the framework to saying they are provided with sufficient funds to work in the interests of society.

What happens to the organisation's health if it is not economically healthy?

How do, for example, ethics and environment affect economic health?

Efficient Processes

Efficient Processes ensure that value is being added at all phases during required work, and that unnecessary costs that would harm the economic health of the company are not being incurred. Working efficiently entails ensuring that all equipment is operating at optimal levels; that people are clear about their focus and responsibilities and the manner in which they are all interdependent upon one another; that the company is working hand-in-hand with suppliers and partners, ensuring that components and support services are available as needed. Working efficiently does not entail employees 'producing' pre-defined quantities of work in specific time frames. This would be Taylorism, which is exploitational in nature. The distinction is essential.

Which role does the discernment between efficiency and effectiveness play in organisation health?

How would deficiencies in psychological health negatively impact process efficiency and the economic health of the organisation?

Employee Physical Health

When we speak of health, it is both natural and important that physical health is generally the first thing that comes to mind for most people. Anybody who has ever had an illness or physical problem, such as a bad back or sports injury, is aware how it directly affects our ability to function freely. Each one of us is a whole human being, not a collection of the components of our body.
And so, we see the essential connection between physical health and psychological health, with the latter being core to well-being.

What happens when ... People feel there is meaning in their work? When they feel adequately rewarded? When they feel valued for who they are, not merely for what they do, for which they are also recognised? When they are engaged with one another and in their work, ensuring the success of the company as a whole?

I imagine many readers also know people who have become physically ill for psychological reasons. It doesn't require an economist or a businessperson to recognise the negative affect of employee sickness on the economic health, product and process quality of the organisation.

Qs

How do ethical and environmental standards affect the quality of internal and external relationships and the economic health of the organisation?

How does personal health affect the quality of service to customers?

Amanda was working as an English teacher in Northern Thailand when she told me her story. She had previously been a successful salesperson in England, driving a sports car, living the good life. One summer's day, while stuck in a traffic jam, she broke town and the tears flowed, as she asked herself the Question that so many of us have asked in life: *"Why am I doing this to myself?"* For Amanda, the work itself wasn't a problem, it was the stress ... caused by continuous pressure combined with the lack of meaning in her work. She sold everything and went to live in Thailand.

Psychological Health

An organisation is a community,
a social space where people work.

The core of psychological health is a sense of belonging. Human beings are social creatures (Aristotle: *"Man is a social animal."*). Human beings need community, whichever form this may take. Belonging incorporates feeling valued as a person, not merely for one's contributions. In healthy organisations, recognition, acceptance as a valued member of the community, is explicit on the part of the leadership as well as the group. There exists a feeling of freedom within the boundaries of the organisation's collective responsibility and interdependency, as well as being a part of something bigger than one's immediate environment. Meaning, a feeling of personal connection with the environment and outcomes of one's work, is an essential contributor to psychological health. When a person attributes meaning to their work, they are far more likely to ensure that their activities are efficient and adding value. They will enjoy their interactions internally and with external partners including suppliers and customers. They will identify with the product and the ethical standards within which the organisation operates.

What is the economic cost to an organisation of psychological illness due to manipulative work practices?

How do poor supplier and customer relationships contribute to psychological stress among employees?

This Sales Director presented his dilemma during a coaching session.

The new CEO is under pressure to increase sales significantly and has informed the successful Sales Director that past success is not relevant to the future. The Sales Director is thinking of leaving.

What needs to change? How?

Quality Products

Whether the product of the organisation is a physical entity or a service, it needs to be fit for purpose, otherwise nobody will buy it. Customers need to consider the price and associated services reasonable, which, unless we ignore ethics altogether, could not be the case without good relationships with suppliers. Ideally, they will also enjoy doing business with the organisation. In the 21st century, a 'quality' specification may well include assurances on the sources of all components used in the production process, the organisation's reputation regarding its treatment of employees, environmental measures.

To what extent can a product whose components are sourced through exploitation, be considered a "quality product"?

How will customer and media complaints about the quality of a company's products affect the psychological well-being of employees?

Ethical Products

What makes a product ethical or otherwise is a multifaceted issue. Questions include the purpose of the product; the intention of the customer(s) in purchasing the product; the ethics of the producer; the ethics of the sourcing of any components, which would include environmental issues, treatment of employees, treatment of suppliers.

How does an organisation's ethics affect its ability to attract well-qualified innovative employees with integrity?

To what extent do the ethics of a company's products influence its customers' perceptions and economic health?

Healthy Customer Relationships

When faced with a choice of alternative products that are similar in nature and function, some customers may simply choose the cheapest. Efficient processes help to keep costs under control. Decision-making processes of well-informed customers are likely to be more refined. For technical products, reliability plays a significant role, whereby quality of service and ease of doing business are key factors. For foodstuffs and clothes, quality, ethics and environment will be considerations. For employees to engage positively with customers, they need to feel valued by their employer.

Qs

What is the economic and psychological impact on the company and its employees of losing customers?

To what extent can we consider a relationship with unethical customers as 'healthy'?

In 2015, millions of refugees were fleeing war and persecution to seek asylum in the European Union. Two German companies refused requests to supply life-threatening razor wire to the Hungarian government that was erecting a fence on their border with Serbia.

Healthy Supplier Relationships

What does it take for a supplier to engage with their customers, rather than simply to provide a product under agreed conditions? Perhaps it is something as simple as feeling valued and being treated like a partner, which of course, will affect the manner in which suppliers react to special requests arising from unforeseen circumstances, delivery date integrity, price negotiations, and more … all of which directly affect the economic health of the company including its ability to service customer orders.

Qs

Many global industries demand that suppliers set up facilities in the vicinity of their production sites. A dependency relationship emerges. To what extent can the customer and supplier be considered healthy organisations?

What is the effect of a dependency relationship on the psychological health of employees in both organisations?

In the years immediately following the fall of the Berlin wall, demand for electronic goods in Germany surged, so much so that one of Europe's most successful mail order corporations seriously misjudged the demand for its showpiece own label video recorder. A simple yet accurate description of their business model was: purchase cheaply, sell more cheaply than high street outlets. When they approached the South Korean white label supplier to request additional units quickly, they were turned down. The supplier gave priority to those customers that treated them as partners and offered them fair prices. Attempts at mediation failed.

Healthy Leadership

A couple of years ago, while preparing a client proposal for a leadership development programme, I conducted a simple online search on "leadership". Google turned up over 800 million hits; Amazon produced 190,000 results for leadership books. In a world of 7.5 billion people, this is perhaps not surprising.

So, what constitutes "Healthy Leadership"? I would suggest that in any (re-)shaping process, each organisation would need to outline the important characteristics for itself.

However, the core to leadership health would surely include conscious ethical standards, viewing and working with people as humans, the ability to let go and also say "I don't know"; as well as conventional leadership skills such as the ability to see the big picture, recognise strategic consequences of decisions, business nous.

The effect of healthy or unhealthy leadership on psychological health, product quality, engagement with customers and suppliers, and ultimately financial results is undeniable.

Qs

The Corona-virus crisis is having a negative economic impact on millions of companies and organisations throughout the world. Some companies have furloughed their staff; others have laid them off. Others still are absorbing the costs and maintaining full employment.

How will these differences in approach affect the psychological and physical health of employees?

How will it affect employee engagement?
How will it affect product and service quality?
What will be the long-term effect on economic health?

Environmental Health

If the environment is unhealthy, the people who live in the environment cannot be healthy. "Environment" may refer to the physical environment (air, land, water, etc), the social environment (organisation, community, country), the buildings or location in which people work. Anything we do affects our physical and social environment in some way. A healthy organisation will ensure that it contributes to cultivating healthy social environments. A healthy organisation will do its best to minimise any negative impact on the physical environment, either direct or indirect, being conscious of how it sources its products and services. Customer perceptions, employee perceptions and directly or indirectly, the finances of the organisation will be influenced by the company's activities.

How does the physical workplace environment affect psychological health affect process efficiency affect economic health of the organisation?

How does the environmental impact of production processes affect customer perceptions and relationships, community health, ethical health?

Smartphones have become an integral part of our lives. In addition to convenience, they have opened up numerous communication and economic development possibilities for people all over the world. They are particularly important for ... poorer farmers in remote regions, disadvantaged groups, refugees, people suffering human rights abuse.

A challenge: How can we manufacture smartphones without environmental or human abuse, while maintaining quality technological standards?

Community Health

An organisation is a community. Simultaneously, it is part of a wider community of families, other businesses and the environment of the immediate vicinity. How is the organisation consciously contributing to the positive development of the wider community? How does this affect the psychological well-being of employees and their families; the perceptions of the organisation in the immediate and broader communities; and hence customers and suppliers?

Qs

How is the local environment affected by the organisation's presence and activities?

How does the organisation's reputation in the local community affect the psychological health of employees, customer perceptions and its economic well-being?

Political Health

Participation in society implies participation in politics. The manner in which the organisation engages with the community, authorities, politicians and business partners is a reflection of its ethical health and will itself be reflected in its results, as well as its relationships with employees, customers and suppliers.

Qs

To what extent can corporate financial contributions to political parties be considered ethical and how might these influence the quality of customer relationships?

How does the organisation's contribution to its local, regional and national communities influence employee engagement, therefore process efficiency, therefore its economic health?

And so … we see … the health of an organisation can only be realistically viewed through the lens of the whole entity.
While each individual characteristic is a highly significant indicator of health, it only possesses meaning and makes sense in the context of its relationships with all other indicators.

All examples, as well as the Questions posed above, serve as illustrations of the relational nature of the organisation system. They constitute but a small selection of those that could have been offered for consideration.

The nature of life in organisations determines that details of the individual characteristics and thus their relationships with one another are continuously shifting.

The interplay among them is intricate and constant. We cannot meaningfully separate any of them.

It is very difficult to provide a clear accurate diagrammatic representation of the dynamic flows within the organisation.

So … let's look at the cloud constellation at two moments in time, with each picture representing a momentary snapshot. When we compare the two images, we can see that some elements have moved more than others.
What has happened?
Who really understands the mysteries of the magic that organisms weave within their system?

Characteristics of a Healthy Organisation
Snapshot 1

Characteristics of a Healthy Organisation
Snapshot 2

Every Change changes the Change.

Your organisation is full of people weaving their magic …
continuously … through time.

Ensuring continuous individual personal health is never-ending work: eating healthy food, exercising regularly, maintaining good relationships, meaning in life.

Ensuring continuous organisation health is never-ending work.
It is not only possible; I suggest that it is a self-generating prerequisite for an organisation to thrive.

Further Questions for Reflection ...

1. Are we setting the bar too high here? Are we being realistic in demanding that all these considerations are embedded within an organisation?

Perhaps. But ... to what extent can life be meaningful without them?

2. Does it even make sense to speak of a "Healthy Organisation" within the extremely unhealthy environmental, social, political and economic system in which we live?

The Question is important and demands consideration.
Yes. It absolutely makes sense to do so. Our systems are interdependent on one another, not independent. Changes in one system will influence other systems.
We have the responsibility to begin with ourselves.

If not here, then where? We start where we can.

3. Have you ever experienced an organisation in which all these criteria are clearly observable?

No. I haven't.
And this is precisely the reason that it is even more important to begin work without delay.

A brief look at the state of our world today ought to be sufficient to highlight the need for a complete reconsideration of the way we live and work.

If we don't start now, then when?

The Business Case for Organisation Health

"What's the business case for this?" is a common Question whenever new ideas are presented to organisation leaders.
In the context of modern-day organisation life, the Question is important. Corporations and companies need to be profitable; organisations and institutions need to work within a budget.

For anybody who doubts the business case for organisation health, I would invite them to reflect on and answer the following Questions:

Last year in your organisation ...

- *How many sick days were lost?*
- *How many days were employees absent due to psychologically-related health issues?*
- *How many days were lost due to personal conflicts among colleagues?*
- *What was the cost (internal and external) of legal procedures due to personal and professional conflicts within the organisation?*
- *What was the engagement level of employees in your company?*
- *What was the sum of penalty payments to customers due to a failure to meet agreed deadlines?*
- *Which percentage of products did not meet quality control standards?*
- *Which percentage of customer deliveries were returned due to quality issues?*

All the above costs constitute wastage. All are directly measurable.
It would be relatively easy to do the maths.

Even allowing for some inaccuracies in reporting …
Even allowing for some sick days being inevitable …
Even allowing for some customers inevitably taking unfair advantage of return policies …
… the additional unnecessary costs incurred by an unhealthy organisation will be significant.

When making the business case for organisation health, these additional Questions are also highly relevant …

- *What was the effect of employees' psychologically-related absenteeism on the effectiveness of their colleagues' work?*

- *What was the cost to potential process and product innovation of personal conflicts among employees?*

- *What was the cost of wasted potential due to lack of or limited employee engagement?*

- *How many highly-skilled employees have resigned due to the business ethics or environmental policies of your organisation?*

All the above costs are very difficult if not impossible to quantify accurately.
They will, nevertheless, always be significant.

During the past 10 years, there has been a 40% increase in workplace psychological illness in Germany. Whatever the limitations of statistics, and allowing for the fact that all statistics are open to manipulation, the number is disturbing. This piece of news was tucked away as a minor item on the website of a serious national newspaper, which could perhaps also be interpreted as an indication of systemic disturbance. I remember attending a conference in 2011 during which the core topic was stress-related illness. The problem is widespread and can barely be considered as "news".

A number is a number. It is real, yet abstract. It is a piece of hard cold albeit important information.
Behind any number are thousands of humans who are currently unable to live quality fulfilling lives. Behind these thousands of humans are many more thousands of humans, in particular family and close friends, people whose lives are directly affected by the illness of those they care about.
Here, we have merely an initial indication of some of the direct social costs of unhealthy organisations. We haven't yet even touched upon the effect on the school teachers of the children whose parents are suffering, or on the colleagues of their partners and friends, or …

Can you remember the day you started a new job or joined an organisation?

If you are anything like me, you may even have been full of enthusiasm and thoroughly looking forward to contributing and being a part of something worthwhile?

And can you also remember the occasions when you were brought back down to earth with a thud, as you discovered the presence of snake-pit politics and the organisation's "establishment", both formal and informal, who informed you clearly how things need to be done 'here'?

Perhaps you can remember more than one such occasion.

Gallup's European employee surveys during the 2010s regularly reported engagement levels as low as 14%.

Potential:

What would be the point of an organisation without potential?

Following a keynote talk at an international conference some years ago, I was challenged to summarise the three key learnings of my work over the past 40 years. Here is the result:

- *People WANT TO contribute.*
  They are looking for opportunities to engage positively ...
- *~~Most~~ All organisations contain massive untapped potential ...*
  resources which they are frequently unaware of ...
- *Difference is a Natural Resource ...*

These principles are embedded into the ideas presented here.

It is impossible to accurately measure the cost of lost potential due to organisation health issues. Regardless of the methods used to arrive at an estimate, the amount computed is bound to be considerable.

The "business case" for Shaping a Healthy Organisation is irrefutable.

Business leaders are ultimately responsible
for the organisation's results.

...

...

Failure to consciously work on Cultivating a Healthy Organisation
constitutes irresponsible leadership behaviour.

How does life come alive inside the organisms
that are organisations, inside communities?

What happens there?
How does it happen?
Why does it happen the way it happens?

...

...

...

Culture ...
... the patterns that influence interactions ...
... the interactions that themselves influence the culture ...

...

...

...

Little is clear ... and herein lie the richness and opportunities of life ...

Don't compromise yourself.
It's all you've got.

(Janis Joplin)

Everyday Life in Organisations

- ***Organisation Culture Alive***
- ***Multiple Lenses***
- ***History***
- ***Daily Interactions***

Organisation Culture Alive

New is never totally "new".
There's always something there.

I remember the conversation well. It was back in 1970 or '71. In a small group, we were working on a school economics assignment in which we were challenged to draw up a concept to improve the condition of public and private housing in the area of North London where we all lived. One of my classmates said, in all seriousness, "Just tear it all down and start again." Clearly, making hundreds of thousands of people homeless was not a serious option. We humoured him and continued.

Whatever we are doing, there is always something there. And this "something" forms a significant part of the context for beginning any intentional organisation culture change work, even shaping new organisations.

Culture, like change, is a widely misunderstood notion.
Far too often, it is presented as the icons that symbolise what a group purports to stand for. Such icons include flags, emblems, mottos, anthems, logos, buildings, professional standards "authorities", national borders. Icons are merely icons, constructed, not natural. Icons are not culture.

Frequently, culture is reduced to the notions of nationality, belief systems, ethnicity, gender, generations. Both national and organisation culture are frequently further reduced to extremely simple behavioural ideas such as "the way things are done around here". While these notions are certainly influencers of culture in any specific environment, they do not constitute culture. Behaviour reflects culture, as well as influencing it.

Behaviour is not culture.

When we consider the complexity of the challenges we face
in making sense of our world, the attraction of simple explanations
is understandable.
In organic environments, they tend to be misleading.

The public works project, like so many such projects, was in serious danger of not meeting its completion date, with the resulting threat of political embarrassment and economically painful penalty clauses. The international consortium was in conflict.
I was sent in as a mediator with the instruction, "Get the facility open!" We did.
Shortly after returning home, a colleague asked me whether the mix of nationalities in the consortium were a cause of the conflict. They weren't.
The critical factors were certainly cultural, yet of a completely different nature: professional and political.

Culture is relational.
Fixed definitions are inappropriate. The dynamic is highly-complex,
evolving through the multitude of interactions whose roots lie in
interactions that go way back in history; roots that play a formative
role in how we think and feel, act and react.
These roots remain, as new interactions through time add to the
potpourri of influencers.
Patterns form.
As we learn and evolve, patterns change.

Effective intentional positive Organisation Culture Change entails
changing the DNA of the organisation.
This is deep work, discovering patterns that influence the roots; with
each member of the organisation encountering and confronting their

own patterns … together, in relationship …; before consciously cultivating the new patterns in the Healthy Organisation, the community, they all wish to be working in.

Unsurprisingly, people who have been through such a process invariably like to talk about it.
The vast majority describe it as intensive, frequently intense, rich and enjoyable.

Clarifying the Nature of Culture

Why is culture so important?
Quite simply, it is culture that determines and reflects how we come alive, how we bring our souls to this entity we call life. On the one hand, it influences life in our communities. On the other hand, life in our communities influences the culture within.
Is this a contradiction? … Not at all. Culture's dynamic specifies that it is continuously changing. Without this natural dynamic, there would be no life, only death … either slow or fast. Culture is aliveness.

Picture a beach ball full of numerous different particles, spinning uncontrollably in a strong wind. The manner in which the particles bump into one another (interact) affects everything else within this environment, which we can refer to as a system. The momentary nature of each of the particles is dependent upon the momentary nature of each one of the other particles, demonstrating the interdependence of everything within the system. Each collision changes the nature of the particles.
Now, let's assume that each particle represents a person. And here we see culture in organisations at work. Culture influences and is influenced by everything within the system: interactions as well as environmental factors. When we consider these bumps as interactions in a social system (a community, an organisation) we can picture the dynamic nature of culture in that social system.

Culture influences and is influenced
by how we feel in any specific environment.
What is our gut telling us about this space?

Culture is the Soul of the Organisation.

We use the term culture to describe the interactions in a social system, a system that is alive. Culture is also the collective term for the creative arts of literature, painting, sculpting, poetry, acting and more … all of which reflect aliveness not only on the part of the creators, but also within the communities in which they live, communities that stimulate them and which they simultaneously stimulate.

Is this a coincidence? I think not. There is no separation.
The etymology of the word 'culture' is the Latin 'cultura' meaning grow, cultivate, care.

Shaping a Healthy Organisation involves cultivation and care.
If the members of the organisation are not doing this, who will?

Culture in a Nutshell

Culture:

The continuously evolving dynamic interaction
of the mindsets and gutsets
of all the actors in the system.

This is not a 'definition' of culture, but rather an explanation.
As a dynamic entity, culture defies definition. We cannot realistically say, "Culture is …".

Anthropologist, Gregory Bateson, famously asked the Question:

"What is the pattern that connects?".

Patterns in organisations are always changing;
and are invariably invisible.

Nevertheless, the patterns that determine and envelop the interactions of an organisation, its choice of customers and suppliers, its relationship to its members and their families, and local communities; the patterns that describe its culture … can, if only to a degree, be laid open.

Discovering patterns will help members understand the nature of the organisation's culture, their part within it, and thereby, indicate the work they need to do to make a conscious shift to cultivate the healthy environment everyone desires.

A Dance …
Culture and Change:
Perspectives through Multiple Changing Lenses

Think of the atmosphere as you walked in the park on a hot summer's day.
That's culture.

Think of how you felt as you were invited into a stranger's home
to share a pot of tea.
That's culture.

Think of your thoughts as you observed the behaviour of the security
and sanitation staff at your nearest airport.
That's culture.

Think of the religious ceremony you observed, and your thoughts
while observing.
Think of the variety of possible reasons that each participant in the
ceremony is attending.
That's culture.

Think of how each participant in the above ceremony might react if you,
a stranger, asked to hold a conversation with them.
That's culture.

Think of the atmosphere in the lively market you visited during your last
trip to Asia.
That's culture.

Think of how you felt and what you observed as you entered your place
of work this morning.
That's culture.

Think of that phone conversation with a colleague who didn't like your approach to tackling a problem.
That's culture.

Think of your thoughts as you observed women doing heavy labouring on the construction site in India.
That's culture.

Think of how your children greeted you when you arrived home yesterday evening.
That's culture.

Think of that politician's speech you listened to on yesterday's evening news.
That's culture.

Think of that time you observed the police in the street apprehending somebody.
That's culture.

Think of the long walk you have to take to the gate in order to enter the beautiful park whose upkeep is financed by your tax contributions.
That's culture.
...
...
Now ... think of the same scene on a different day, reflecting how you felt, what you observed, how you felt about what you observed on this different day.
That's culture.
...
...

Now ... observe yourself thinking and seeing and feeling these thoughts and scenes and feelings ... and ask yourself ...
Why ... am I thinking/seeing/sensing what I do?
That's culture.
...
...
Now ... slow down ... perhaps close your eyes ... and imagine each of these scenes again, with the other person/people observing you ...
That's culture.
...
...
Culture is so multi-facetted, nebulous in nature. Yet, above all, it's alive. It is what you think and feel about what you observe. It changes ... continuously. It has multiple influencers that themselves change, as does the manner in which they influence. It's messy, very messy ... The inherent richness and beauty in life's messiness ...
And of course, your perspective is only your perspective ... through your lens ... in a particular moment.
Others will observe the same event and see/feel/sense differently.
They each have their own lens.
This is culture.

Culture has no categories; no boundaries. It changes and evolves at every moment. It is so difficult to pinpoint yet it has such a strong presence. It lies at the core of our lives, how we perceive absolutely everything in the world. It is the soul of society, organisations, families, communities ... experienced differently by each member of these groups ... transcending contexts.

Now ...
Think of the difference between your home city today
and 20/30/40 years ago.
That's culture change.

Think of the environment in any office or factory today and compare it
with real stories of the same place from 20/30/40 years ago.
That's culture change.
...
...
Think of how you see the world differently today
compared with 10/20/30 years ago.
Without doubt, you've learned a lot in that time.
That's change.

As cultural patterns change, the nature of the roots around
which we shape our societies and communities, change.
...
...
Culture Changes ... continuously.
Being alive with culture as it changes, is a dance.

The Inevitable Messiness of History

"I start in the middle of a sentence
and move both directions at once."
(John Coltrane)

Context matters … Always.

Life in organisations seldom, if ever, reflects the neat diagrams that illustrate their structures. Everyday life tends to be more nuanced. Only rarely is anything crystal clear.

Cultivating a Healthy Organisation … is a messy business.

Where should we start to work?

Let's start at the beginning …

Where else? You might ask.
Well … as we have already seen, the beginning is rarely the beginning.
In fact, it's always somewhere in the middle.

We start where we are.

... history ...

There is always history.
Usually, when leaders start organisation change initiatives, they are working with an existing entity. The organisation's history is integral to its being.
Even when we are working with completely new entities, there is still history ... the history of all the people in their life to date.

History is always embedded within the present ...
... intertwined with and shaping our sense-making of our experiences in life, our families, culture, traumas ... all of which shape us as we simultaneously shape our environments ...

Whereas it is relatively easy to define and lay out clear processes for functional mechanical work in organisations, working with the people to intentionally shape a working environment entails encountering living organisms in their messy often undefinable continuously changing world.

There are no neat models which can provide a realistic framework.
There is no formula.
There is no blueprint.
Each situation is unique.
...
...
We never "know" what is going to happen.
We have to learn to use and trust our intuition.
Wisdom is readily available in every system ...

To which music will you be dancing in your organisation?

... At times, it may resemble rock 'n roll.
... At others, it may be a smooth waltz.
... At others, it may feel like an irritating staccato.
... It may even seem like a group of people trying to dance in a muddy field while it's pouring with rain.

Where do we find orientation?

During any intentional organisation change initiative, one of the primary responsibilities of the nominal leaders and any external consultants working with them is to provide orientation and support within this muddled arena.

How? You may justifiably ask …
We will provide an answer in the pages that follow.

We are using our intuition.
We don't always know what we're doing.
Even when we do, we're still experimenting.
Each situation is unique.

"I've found you've got to look back
at the old things and see them in a new light."
(John Coltrane)

Ensuring Healthy Organisation Interactions

For the vast majority of organisations anywhere in the world, cultivating a healthy environment is, practically speaking, a Culture Change Initiative.

We see and sense the organisation's culture through the quality of interactions in its system.

These interactions are simultaneously a reflection and an influencer of its culture.
Imagine the organisation as a dance floor. The dances, the movements occur as a continuous flow … a flow in which all the dancers, individually and as a group, consciously and subconsciously, offer to and elicit from the space.

Which interactions in the organisation are pivotal for shaping its culture? Precisely where do we start? How do we start?

We start right in the centre. There is nowhere else.
We start by working on and with all interactions simultaneously. There is no other way. None is an independent entity. Successfully working with living systems does not lend itself to lists and priorities. We have already observed their interrelational nature.

Nevertheless, we can perceive each of the descriptors in the diagram as a Question to be answered by members of the organisation.
Not a Question to be answered independently, rather in relation to … every person involved, the business, all other Questions, … life in the organisation.

The following chart presents an overview of key interactions.
It is incomplete: any chart would be.

Organisation Interactions that Reflect its Life and Culture

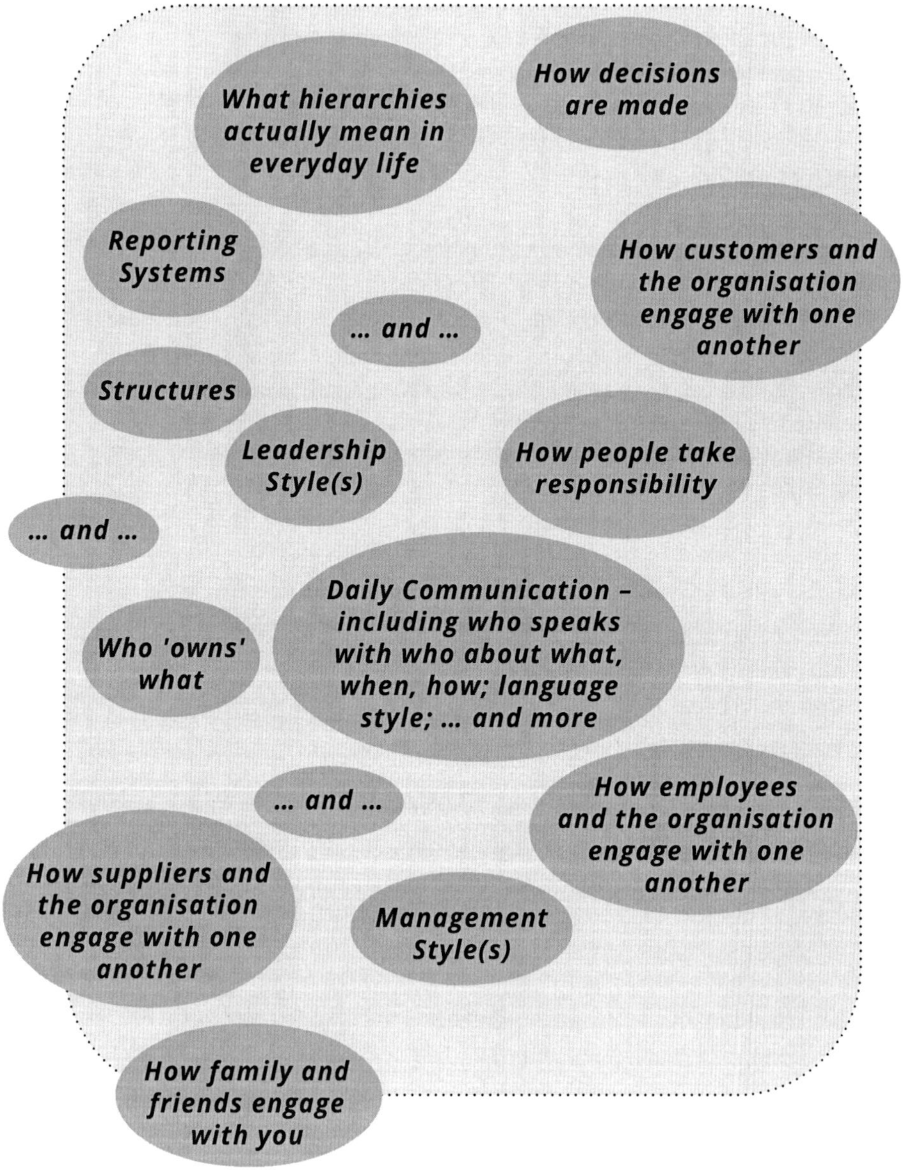

Who engages in the work of cultivating these interactions?
The people in the organisation … everyone. Nobody else can.
They work together … in interaction.

Let's take a brief look at these key interactions and how the quality embodied therein influences the whole organisation.

We need **Structure.** Perhaps paradoxically, it provides freedom, outlining the framework of the space in which the organisation's members work … or maybe we should say, dance. Core Questions concerning organisation structures include what they are designed to do; which forms of communication they support and/or hinder; how they can be set up so that they support the process of continuously generating the desired environment.

Hierarchies. It is not uncommon to hear people talk of organisations without hierarchies, as though their existence is in some way negative. There is always hierarchy. We find it throughout nature. Every organisation too, has its hierarchy, be it formal or/and informal. The significant Question is not the binary, should or shouldn't it exist; but rather what the hierarchy actually means in practice. The hierarchical principles of the organisation will be reflected in its decision-making practices, leadership and management styles, **Communication Channels, Communication Style,** reporting systems … and more.

Organisations need **Reporting Systems.** After all, those legally responsible need to know much of what is going on, if not all details. Questions for designing an organisation's reporting include … the appropriate channels, degree of detail, timing (stage of work) of reports, the nature of a "report", ethical principles.

While the organisation's formal policy is important, it may or may not be directly relevant to the quality of life within. Indeed, it is the daily practice of **Leadership** that to a large extent determines the quality of interactions in a company, including … how employees engage with one another, their work, the organisation, more succinctly, their sense of **Ownership** and **Responsibility** for their work; communication styles and channel; how challenges and problems are met; information flow; **Decision-Making** approaches; attitude to external stakeholders … indeed, everything.

During the re-engineering fad of the 1990s, layers of management in many organisations simply disappeared … until senior management realised the error of their ways and stopped listening to those consultants who were selling snake oil packaged as vitamins. Organisations require **Management.** Tasks, processes, administration, government regulations, logistics, projects and more, all need to be managed. Do people need management? As such, perhaps not. Aspects of their work, however, do … including, their work-loads, working environment and access to resources. When the principles involved in managing administration, processes and projects spill over into attempts to manage people, problems are likely to arise. Healthy Organisations make the distinction.

Organisations regularly refer to the importance of their stakeholders, in both meaningful and disingenuous communiqués. Most of the time, "stakeholders" refers to those with a direct financial interest in the business, specifically, creditors, directors, employees, government (and its agencies), owners (shareholders). This use of the term is problematical, as it tends to view the organisation as an independent entity, implying that there are people and groups who are not stakeholders of an organisation.

The existence of an organisation as an entity and quality of life within, are invariably related to all other entities within the organisation's eco-system. They are not independent of one another. Every individual, every organisation and community within this eco-system is a stakeholder. Obvious ones include … **Customers, Suppliers,** employees' **Family** members, close friends of employees, the local community at all locations.

A Healthy Organisation will ensure that good quality relationships with all stakeholders are continuously nurtured. These people are partners more than external "stakeholders".
A positive or negative relationship will impact other interactions within the organisation positively or negatively.

And here we have a clear illustration that ...

The boundaries of the organisation
are not the boundaries of its system.

Healthy Organisations cultivate and nurture their systems.

In the diagram, **"... and ..."** represents additional key interactions that need to be addressed and nurtured in every organisation. The context determines the specifics.

Cultivating healthy interactions entails creating a space for conversations that directly and indirectly address these Questions. A picture of the dance emerges, together with a recognition that the responsibility for fostering and nurturing health is collective and continuous.

New patterns evolve.

Organisation Interactions that Reflect its Life and Culture

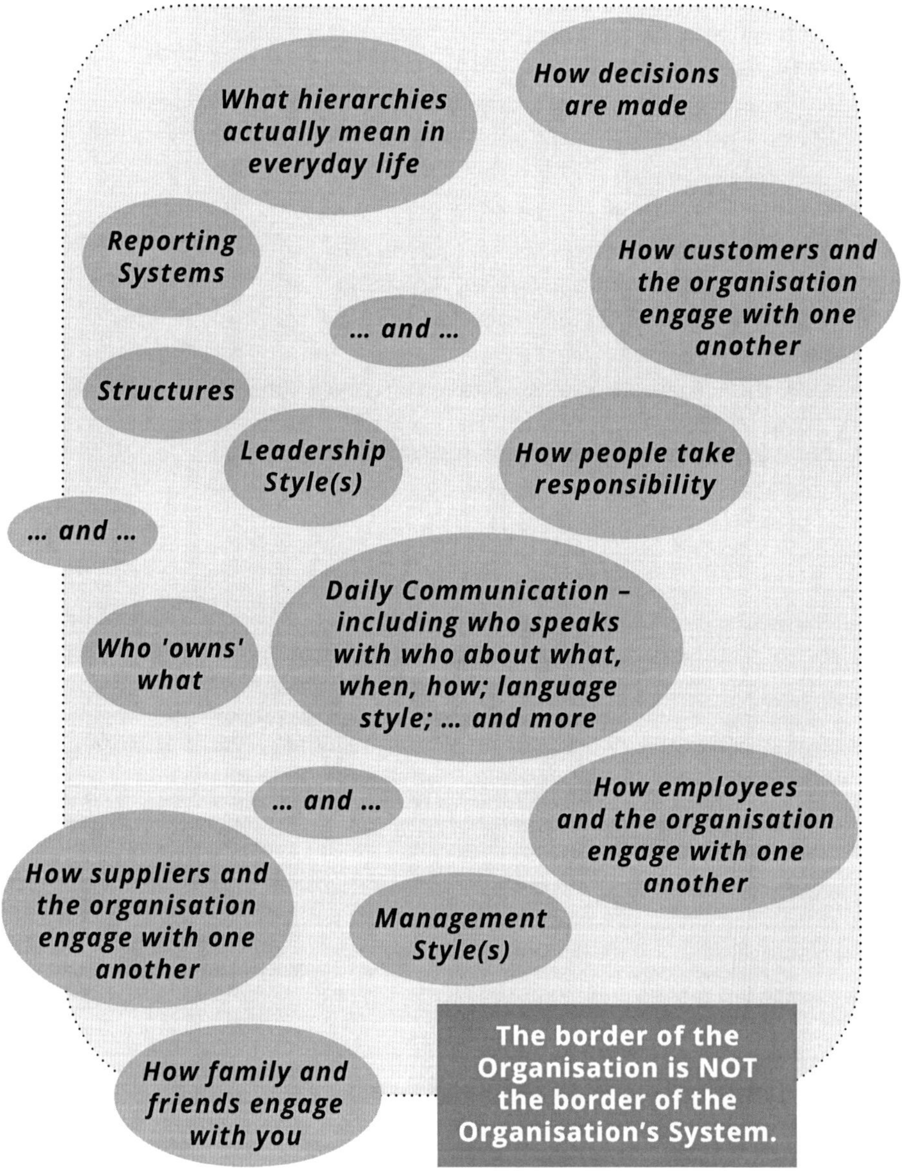

"We see with our legs"

(Heinz von Foerster)

Cultivating
Healthy Organisations

A Practical Framework

How can we Cultivate a Healthy Organisation?
A Practical Framework

Let's summarise our insights to date about Healthy Organisations …

We have posed the fundamental Question for any organisation that wishes to thrive.
We have outlined the elementary characteristics of a Healthy Organisation.
We have seen the significant underlying role of history in all interactions.
We have touched upon the key interactions that simultaneously reflect and shape the organisation's culture.
We have ascertained that any meaningful initiative needs to involve all members of the organisation.

And now … perhaps … you are anticipating …

… an outline of that magic formula, that manifesto, that clear model outlining the path with Seven Steps to Heaven, enabling you to miraculously arrive at your desired destination?

Dream on ...

While the dance of the natural interactions of organisms may well be magical, I would like to think we now understand that they neither define nor follow clear paths, especially those paths that are pre-defined by thought.

As I'm writing this passage, I'm listening to Miles Davis weaving his magic on his horn while playing Seven Steps to Heaven. I cannot hear a formula; I cannot perceive a clear path; he is certainly not following any kind of prescribed manifesto. He is bringing to life his intuition and skills, skills learned through decades of practice while sensing and feeling his and others' music, reflecting the complexity of style that makes jazz what it is … magic.

So … let's expand and clarify and specify this crucial and highly complex Question …

… How can we Cultivate a Healthy Organisation? …

… while emphasising the all-important nuance that differentiates a framework from a model …

Qs

How we can incorporate all our insights into the act of Cultivating an Organisation … so that it thrives …
… without falling into the trap of squeezing ourselves into the straight-jacket of pre-fabricated models …
… models that are merely abstractions of an abstraction, models that repeat and reinforce historical pathological patterns?

A lifetime of working both in and with organisations has taught me the benefits of working with frameworks that simultaneously …

… create an open space which enables the emergence of new generative approaches for overcoming challenges …

… while providing the orientation, focus and structure necessary for holding and shaping the potential of a broad range of possibilities.

What follows is ...

... A Practical Framework for Cultivating Healthy Organisations ...

The Hodja always wanted to learn something new, and one day he had a sudden inspiration to learn how to play the lute. He approached a music teacher and asked him, "How much do you charge for private lute lessons?" "Three silver pieces for the first month; after that one silver piece a month." "Oh, great!" exclaimed Hodja. "Then I'll start with the second month." (Attributed to Nasreddin Hodja)

The profound wit of ancient wisdom can teach us so much ...
if we are able to see what we are looking at.

There are no short cuts for deep relational work.

Let's remind ourselves ...

Cultivating a Healthy Organisation entails Organisation Culture Change. Organisation Health is a holistic phenomenon.

Many, possibly the majority of organisation improvement measures focus on developing individual business units or teams. While this work is important, and certainly does influence the culture of the organisation, perhaps even significantly, it does not constitute culture change.

A culture change initiative necessarily involves working with the whole organisation as one entity.

Working with the whole organisation means working with an existing eco-system. There is no new beginning … but of course, there is … the beginning of a conscious initiative … which itself incorporates several traps as well as a wealth of opportunities.

I suggest using the framework outlined here as a kind of rudimentary yet clear orientation.

While any of these core principles alone may well help any organisation to improve its environment and results, it is their combination that ultimately enables an intentional change initiative to become successful.

Due to the unique conditions and circumstances in each organisation, as well as the working preferences of each facilitator, I consider it unwise to offer prescriptive advice on how to bring these principles to life.

Each leader and facilitator need to define the "how to", on each occasion, together with the groups they are working with.

Cultivating a Healthy Organisation Principles Underlying Successful Organisation Culture Change Initiatives

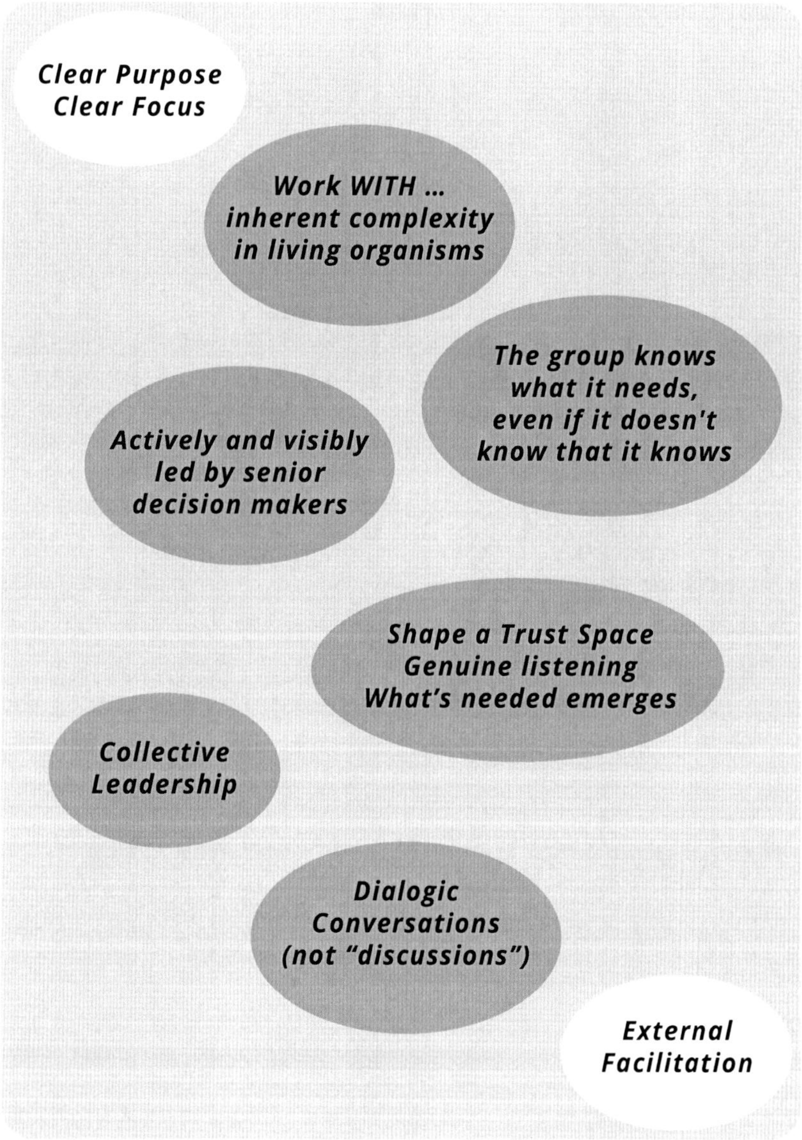

Avoiding the Blinding Reflections of Project Glasses

Conventionally, new initiatives in organisations tend to be structured in the form of projects or processes. Objectives will be attributed and these need to be met, after which the initiative is considered "done" and may well be used as a stepping stone to the next one.

While well suited to projects, such an approach is completely inappropriate for any culture change initiative and certainly not for any work in which the core focus is organisation health.

Why?

Firstly …
When it comes to health and culture, notions that describe dynamic continuously changing conditions in any environment …
… there's no such thing as "done".

Secondly …
Organisation Culture Change is relational work.
Relational work is never "done" … perhaps not even after an organism has died.

The work of shaping, cultivating and maintaining
a healthy culture is continuous.
It never ends.

Relationships give a living system its integrity.
We work in relationship with relationships.

The Principles

Principle: Clear Purpose. Clear Focus.

Without orientation, we're lost.
Purpose provides orientation.
Focus provides a navigation aid.

Here, we immediately bring the fundamental Question to life:

"What do we want our Organisation to BE?"

The organisation's leaders need to be perfectly clear about ***why*** they need this initiative, ***why*** they want to start this initiative ***now.***

Employees and managers will almost certainly be asking these, as well as other related Questions. It is likely that many will have concerns about the meaning of the initiative for their work. There may also be suspicions about whether the initiative incorporates any kind of hidden agenda to reduce jobs. Concerns need to be taken seriously and people need to experience that they and their concerns are important to the senior leadership.

By virtue of their formal role, the senior leadership are responsible for providing orientation, communicating a genuine clear message with appropriate focus and humility. The message does not merely provide orientation, it constitutes the essential invitation to all members of the organisation to actively involve themselves in (re-)shaping the organisation in a manner that they feel to be appropriate.

A word of caution regarding "purpose" …
"Purpose" is simultaneously important and a trap.
Let's return to languaging …
When misunderstood, used, or more accurately, abused in the sense of defining purpose as a "thing" that we "ought to" have or find … it is pure nonsense.

Used in the sense of "meaning in life", it remains core to our sense of being.

Used in the sense of "the reason", it provides valuable orientation.

While people always require orientation, so much of which comes from our inner compass, we need to be wary of the danger of extracting the purpose of an initiative from its environment, its ecosystem. It is here that so many initiatives fail because they fail to take the intricacies of the relationships in the organism(s) into account.

In his classic work, Steps to an Ecology of Mind, Gregory Bateson warned of the trap of Conscious Purpose: *"If you allow purpose to organize that which comes under your conscious inspection, what you will get is a bag of tricks - some of them very valuable tricks. […] But still we do not know two-penn'orth, really, about the total network system. […] Wisdom I take to be the knowledge of the larger interactive system - that system which, if disturbed, is likely to generate exponential curves of change."*
He spoke of the arrogance of science in assuming *"…that the individual human being has total power over the system of which he or she is a part."*

The danger of manipulation, conscious or otherwise, is ever-present.

There is wisdom in the system …
Learning to listen to and trust our intuition is so important.

Principle: We Work with … the Inherent Complexity in Living Organisms

What do we mean by … the inherent complexity in living organisms?

Their relationships, their history, their nature, their dynamic … their integrity as living organisms.
Without these, they would be dead not alive.

The notion of complexity merely describes the nature of living organisms.
Nothing is stable, ever, even though it may appear to be.

Every change changes the change.

What does this mean for designing Organisation Culture Change Initiatives?

We recognise that whatever we are defining is changing continuously.

We set aside notions of management, control and planning.
Management, control and planning remain important for operational processes.
Effective operational procedures require predictable outcomes … which is not the case for natural changes in organic environments, including genuine organisation change.

We work with the inherent complexity of the organisation by incorporating open-endedness into diagnoses, intentions, communication, interactions, initiatives.

We explicitly recognise the critical role of the organisation's history in forming and defining its present-day patterns, both visible and hidden. We can only hope to shift these patterns by working with them.
We have neither pre-formulated goals nor specific outcomes in mind.

Consider any conflict, be it economic, business, political, organisational, family ... whatever.

The pathology of the conflict lies in the system in which it is living ... and this is the only space in which it can be overcome.

Principle: Organisation Culture Change Initiatives are actively and visibly led by the Senior Decision-Makers

Why?
By virtue of their position, the behaviour of people who are attributed the title of CXO, Managing Director, or similar, the nominal leaders of the organisation, constitutes the example for everyone else within it.

In practice, this means that what they actually do as well as what others perceive that they do, are critical influencers of all human interactions involving the entity.

The same applies to what leaders say. The integrity of their pronouncements is entwined with people's perceptions of their activities.

"People follow what you do, not what you say," is an oft-repeated sentiment.
For leaders, it's an essential truth.

Actively and visibly leading the initiative does not mean that these leaders take all the significant decisions. If they interpret their role in this way, the initiative will likely fail. They need to demonstrate their trust in the members of the organisation by letting go, while engaging actively with and supporting the work of the initiative.

This is not the same as delegation.
In this context, delegation is not relevant, as it incorporates the notion of bestowal, the leader's right of decision and control. Bestowal and control are contradictory to organisation health.

In parallel, members of the organisation can observe their leaders' willingness to consciously work on their own development.

Principle: Collective Leadership

There exists no part of my personal story that does not influence my interactions with others, and thereby, the group context and its environment.

An organisation is a community in which people work.

How could it possibly thrive without the active involvement of all its members?

Collective Leadership refers to nothing more profound than everybody in the organisation actively taking responsibility … individually and collectively … for all activities and outcomes.

This principle follows nature's relational dynamic.
Individuals influence this system, while themselves being influenced by the dynamics within it.

It is high time for the delusional notion of a disconnected personal responsibility involving individual KPIs, reward and blame, heroic deeds, and more to be consigned to history.

The IT company's structures, systems, and reward systems encouraged internal competition while actually discouraging cooperation required to deliver complex projects to clients in the required quality. During a quarterly leadership team workshop, one participant suggested throwing out individual KPIs and replacing assessments of success and progress with project-oriented appraisals. Everybody immediately supported the idea.

One may ask whether it is not paradoxical to speak of collective leadership having just emphasised the importance of nominal leaders actively leading the initiative.

While this may initially appear to be paradoxical, it is not.
Neither is it in the least contradictory.

Generative culture change towards a Healthy Organisation is not possible without everybody assuming co-ownership. Cultivation of a new soil is inherently, a collective effort.

And so, we have … Collective Leadership … in which the vital challenge entails ensuring, collectively, that this leadership is generative. The signature role of the nominal organisation leaders inevitably remains.

In any living system, an intricately intertwined ecology is at work …

Actively working with the notion of collective leadership as a dynamic may well entail a significant shift in attitude for everybody in the organisation.

In most public institutions, traditional corporations and established family companies, this will almost certainly be the case.

As with any work involving deep personal change, shifts rarely occur overnight.
New patterns need time to take root. Individuals must be willing to work consciously to become aware of their own patterns, enabling an unfolding of new ones.

Patience is necessary, together with approaches that encourage truly dialogic conversations; conversations that open a space for the evolution of deep-level trust within the community.

Any attempt to manage or control the initiative will fail.

People need air to breath.

Orientation and focus are fundamental.
Here lies the role of the external facilitator … with a light yet steady hand.

Principle:
The Group knows what it needs, even if it doesn't know that it knows

The world is swimming, if not drowning in "experts", "consultants" and "leaders" who know what is "best" for any groups they are working with.
And in order to be recognized as true heroes, many come equipped with perfectly rational models that they are ready to implement to achieve "optimal" results. They know.

Except, of course, that these experts don't "know".

Notwithstanding any claims to the contrary, external experts and consultants simply cannot "know" what an organisation needs.

Neither can organisation leaders "know" what is good for their people.

How could they possibly "know"?
Each member of the organisation is an individual, with individual perspectives, sourced from multiple individual collections of experiences. These experiences are expanding during their interactions as a member of this organisation. Each individual is only relevant within the organisation as a member of the group in the context of this organisation.
The system is alive.
The group is far more than the sum of its members.

The group of highly intelligent young people, all intent on making a positive impact in the world, were working on the challenges of formulating a strategy for their organisation. Naturally, there were personality differences; naturally, the challenges contained polarities and paradoxes. These were not in the least problematical.
The group were open, accepting, tolerant and good listeners, working relatively smoothly until suddenly ... they were stuck.
They looked at me. I looked at them, shrugged my shoulders and walked to a far corner of the room.
They continued ... and the path they needed opened up.

They knew what they needed.
They also needed to learn to trust themselves.

Have you ever experienced that feeling of discomfort, a churning, in a situation where you sense that something is not quite right, doesn't quite fit, yet you are unable to pinpoint precisely what this something is?

This has nothing to do with knowledge in the conventional sense; neither should it be mistaken for consciousness. This sense is what I refer to as an "inner knowing"; yet any definition would remain inadequate. It lies beyond concepts of rationality or logic.

I have experienced this feeling at a personal level.
I have also frequently experienced it among groups while facilitating.
You sense that they know what they need, despite the fact that they may appear to be struggling to find it and name it.

The group field will evolve this "what they need" … IF … they have the space, including the time to do so.
Let go. Trust them.

A new Director had recently been promoted internally to lead a business unit in a global logistics corporation. Tony was friendly, open, focused and well-respected. During the first quarterly strategic review following his appointment, the Senior Management Team were struggling: something was bugging him and the group. The whole group sensed it, yet nobody was able to pinpoint the issue.

Tony then expressed the need to experience how others perceived their relationships towards him and one another. Everyone else in the team was enthusiastic.

The knot broke.

The group itself is alive, attempting to thrive.

The group knows what it needs, even if it doesn't know that it knows.

Principle:
Conversations in Dialogue

We are always learning.
Learning in the organism occurs in interaction.
Interactions are the lifeblood of an organisation.
Conversations are the compost that fertilises the soil that constitutes the interactions in organisational life.

What do we mean by "conversations"?
Conversations are the essential medium of interaction among people in any social space.

We conventionally consider conversations to be verbal exchanges between two or more people. While some such exchanges may fit this description, as a definition it is incomplete.

All interactions are conversations …
Shaking hands is a conversation; turning your back on someone is a conversation; failing to return a call is a conversation; failing to keep a commitment is a conversation; failure to admit messing up is a conversation; an acknowledgement is a conversation; a smile is a conversation; a thumbs-up is a conversation … and much more too.

Oral exchanges, which may take on a variety of formats, are also conversations.
Examples include … discussions, debates, arguments, statements, monologues, presentations, dialogues.

And then there is tone of voice; the tone that accompanies any contribution.
A popular German saying: *"The tone makes the music."*

A key challenge for the evolution of a healthy organisation is to enable quality generative interactions … continuously.

As social animals, human beings have an intrinsic need for connection, relation, meaning, being recognised and heard. Such generative interactions form an essential backdrop to the genuine cooperation communities need in order to thrive. It is quality interactions that enable the development of quality products and services in corporations. For innovation, they are essential.

Conversations are relational.
So … let's ask …

Qs

What is the nature of the conversations you wish to be a part of?

What is the nature of the conversations that will cultivate a truly healthy organisation?

Each person will find their own answers to these Questions.

Answering personally …
I would like to engage in conversations …
… that contain and induce inquiry, options and curiosity;
… where all contributions are acknowledged, whether or not we agree with them;
… where reflection is encouraged;
… where Questions are put forward, without necessarily expecting an answer.

Such conversations require presence, presence in the spiritual sense, meaning full attention without concentration. Presence entails letting go of preconceptions … enabling new wisdom to arise (letting come). Conversations of this quality enable an opening to the wisdom inherent in any system.

Conversations in which shared meaning arises in groups are an extremely powerful force, opening fields of genuine potential.

David Bohm (1917-1992), a quantum physicist, gifted us the concept of ***Dialogue,*** one of the most powerful approaches I have experienced for permitting shared meaning to emerge in groups.

The term dialogue is frequently misconstrued to simply mean two or more people having an oral exchange. Literally speaking, this interpretation may well be true.
These exchanges, however, do not describe the nature of dialogic conversations.

Bohmian Dialogue, named after its initiator, involves far more than words.
In David Bohm's own words …

… "Dialogue is the collective way of opening up judgements and assumptions."

… "This will make possible a flow of meaning in the whole group, out of which may emerge some new understanding. […] And this shared meaning is the 'glue' or 'cement' that holds people and societies together."

Lee Nicol, in the foreword to David Bohm's book, On Dialogue, describes Dialogue as follows …

… "It is a process which explores an unusually wide range of human experience: our closely held values, the nature and intensity of human emotions, the patterns of our thought processes, the function of memory, the import of inherited cultural myths, and the manner in which neurophysiology structures moment-to-moment experience."

Dialogue moves beyond the fragmented thought patterns that we conventionally learn to assimilate, towards a space that enables what we might call "collective meaning making".

Is an organisation able to be healthy without this space?
Possibly.

We would be wise never to underestimate the natural influential power of learned ingrained patterns on communication. Together with the range of experiences and emotions that each individual brings to each conversation, it is so easy for one word or one statement to invoke a conflict. After all, so many words can have so many different meanings to so many different people. Associations are powerful emotional prompts.

Dialogue is a practice. Like all deep practices, it takes time to learn and integrate.
Nobody is perfect. There are times when we all inadvertently slip into old learned patterns, especially under duress.
We need patience with ourselves and others.

Two friends, co-owners of a small flourishing catering business found themselves at loggerheads, and requested me to mediate.
Neither was able to coherently formulate their problem, which appeared to be both personal and business-related.
I informed them that I couldn't solve their problem for them, but perhaps they could.

After spending some time on framing, focusing and presencing to enable the opening of a dialogue space, I wrote two Questions on a sheet of paper, placed it on the floor between them, then remained silent:

"What do you really want?"
"Why?"

With time, they opened up towards one another, discovering, among other things, how they had both changed since starting their business.
In the following sessions, they shared how they were able to re-fertilise the field in which they had previously experienced so much success and enjoyment.

For organisations …

For members of an organisation to learn the practice of Dialogue and integrate it into their interactions, will certainly not guarantee the evolution of the thriving organisation they undoubtedly wish for. The chances, however, will increase significantly.

For communities and societies …

If parties involved in violent conflict were able to engage in genuine Dialogue without the interference of external political interests, I suggest many of these conflicts would resolve themselves … as these parties … together … focus on cultivating a healthy society … together.

Principle:
Shape a 'Trust Space' in which people genuinely listen to one another

It doesn't take much effort to encourage people in organisations to perform tasks.
Employing people as "resources" to "perform tasks" is far removed from the notion of health we are considering here.

In a Healthy Organisation, people engage naturally with their work, and positively with colleagues and partners. They feel responsibility and ownership for their activities and outcomes.

Would this be possible without trust: trust in the leaders of the organisation; trust in one another as colleagues; trust in the integrity of the system that is the organisation?
It is possible.
However, I very much doubt whether deep intrinsic engagement is possible without trust.

To trust or not to trust is an essential Question.
Functional trust … the kind we have in a bus driver or a pilot to get us safely to our destinations, or the ability and skills of a colleague to successfully perform a task … is not the issue here.

What is important here is relational trust … incorporating a sense of belief in the integrity of those with whom we interact.

Yet the notion of trust is understood in so many different ways.
Each person enters a space of interaction with their collections of individual experiences.

So, before continuing … I invite you to … PAUSE … and reflect …

Who do you "TRUST"?
How do you "TRUST"?

What needs to happen for people
to bring their whole selves into their work?

Would this be possible without a Trust Space?

A **Trust Space** is a place where you feel completely free to be yourself. There are no borders, no boundaries, no "what-ifs".
You feel safe, supported, protected and accepted, regardless of what you say or do. And from this space you can grow.
For groups, it is the space from which they as a system within a system can grow.
It's an emergent space for ideas, developments, solutions to challenges that people may not even have thought of previously.
A trust space is an emergent arena for Personal and Society Development … a dialogic space.

The **Trust Space** is full of paradoxes:

- It is only a sense, a feeling, yet it is very real for the person experiencing it.
- A Trust Space is always individual – it is after all, an individual feeling. Yet, it only exists in relationship to others.
- Other people in the group may or may not sense the same trust in the group as you.
- A Trust Space is the concrete foundation essential for the deep relationships that enable hidden potential to emerge. Yet, once you have experienced a Trust Space with a group of others, you are likely to sense you have a very delicate plant that needs continuous nurturing. It does!

Within a trust space, there are continuous flows of relationships between the people, experiences and ideas, including those people that may be thought to be outside it, yet are brought in by members of the group. Everything is intermeshed at multiple layers.
This generative evolving flow of systems within systems creates a notion that itself is binary …

Either trust exists in the space or it doesn't. … Another paradox …

If someone says that they trust another person concerning "this" but not "that", can we truly say that there is trust in the space? Any hint, however small, of lack of trust is restrictive. It closes the space of potential.

Experience is the oxygen of the Trust Space.

The Space needs to be opened anew on every single occasion.

A Trust Space is so personal that we can never (and should never) speak for others within it.
A Trust Space can never be forced.

There is no force.
It can never be assumed.
We can only set up the environment that enables it to emerge.

The Trust Space is only a true space if you surrender yourself to it.

Trust is a conversation with oneself … in every single interaction.

Principle:
External Facilitation is a Prerequisite for Success

A traditional, financially healthy, family-owned engineering company with 1200 employees, was struggling to attract qualified engineers. In accordance with traditional management principles, most decisions were taken at senior level. However, younger qualified engineers expected freedom to innovate in their work. The challenge and The Question we were focusing on:

"How can we make ABC Co. an attractive employer
for young engineers?"

Some senior managers were resisting proposals; the less experienced managers were holding back, for fear of compromising their standing within the group. They were stuck ... until I asked the paradoxical Question: *"What will this company be like in 5 years if you do nothing?"*

I wrote the Question on a flip chart and stepped to the back of the room.

The group were blind in the surroundings of the forest they had proudly planted, grown and nurtured. This kind of blindness is a common affliction.
Overcoming it requires a skilled external lens.

What does a Facilitator 'do'?
A facilitator creates and holds the space that enables a group to focus on finding their responses to their challenges.

S/he provides orientation, focus and structure, taking care to limit direct interventions.
Most interventions take the form of Questions, which may be conscious provocations.
Suggestions to the group are genuine invitations.

An intervention does not necessarily need words. It may be a slight movement, a nod, hand signal, or similar. It may also take the form of a silent invitation to form a dialogue circle.
The fewer words a facilitator uses, the better.

Why can a group not facilitate itself?
Simply because the people are intertwined within their system.
Furthermore, they need to focus their energy on their challenges, rather than the dynamics of the group.
Each member of the group is an organism in the system, with a natural interest in its outcomes. Each is biased. Such bias is natural, neither negative nor positive.

Conventional organisational life invariably contains pressures to perform, to achieve, to fit in, to produce, and more, all of which actually close space. Meeting complex challenges requires a space-opener and space-holder to enable the power and potential of a group to come to life.
Here lies the role of the facilitator.

The group is responsible for, and owns all content as well as any outcomes.
Facilitators free themselves of all interest in the outcomes.
A facilitated intervention is not a process.

The group knows what it needs, even if it doesn't know that it knows.
They need a space holder.

True Facilitation is both a Practice and an Art, rooted in self.

Learning the art of facilitation takes years of practice, and the work never ends …

"Water is fluid soft and yielding.
But water will wear away rock, which is rigid and cannot yield."
...
"This is another paradox: what is soft is strong."

(John Heider, The Tao of Leadership)

One herb alone does not make a spicy dish

In the field that constitutes the organisation, introducing any one of these principles into the working framework, contributes positively to the natural fertilisation required for healthy cultivation.

One ... or ... two ... or ... three ... or ... of these principles are not enough.
It is the combination of all ingredients that enable the flows of meaning which communities need in order to thrive.

The more an organisation works with these principles, the more deeply they become embedded in its soil. As they do so, new patterns form, and these too become fluid.

A conventional Question formulated according to mechanical thought, might take the form:

"If we embed these principle in to our Organisation Culture Change Initiative, can you guarantee success?"

"Of course not!", would be the only reasonable response. "Life has no guarantees."

What do you want Your Organisation to BE?

It's up to you, the people of the organisation, to ...
compose the music ... and dance ...

"Others have seen what is and ask why.
I have seen what could be and ask why not."

(Pablo Picasso)

Transitioning to Action

A Reflective Interlude

From

... BEing ...

to

... BEing and DOing ...

Dobedobedo
(Frank Sinatra)

Dancing to Changing Rhythms

And those who were seen dancing
were thought to be insane by those
who could not hear the music.
(Friedrich Nietzsche)

It is always healthy to Question norms and conventions.
We may or may not decide to change them.
Failure to Question them is potentially harmful to individuals
and the societies in which we live.
Not everybody will understand.

We need to Question our narratives about change.
We need to Question our narratives about the purpose and design
of our organisations.
Indeed, I hope it is now clear how essential it is that we change
the way we think about change; the way we work with change;
the way we think about organisations; the way we design our
organisations; the way we develop our organisations.
Not everybody will understand.

Life creates its own rhythms and as these rhythms change,
so does the dance.
The dances within organisations are no different.
Why should they be?

Perhaps you can recall the staccato as you entered the offices
of your new employer for the first time?

How about the rock 'n roll when the project team finally found a way to overcome that irritating technical challenge, and then set about testing their theories again and again?

Or the smoochy ballad as customers looked on in awe while you and your partners presented those new developments that surprised even yourselves?

What about the angry trombone when employees rebelled, openly refusing to continue dealing with that choleric customer?

A healthy organisation environment will find space for a variety of rhythms.
After all, everybody needs freedom to move.

The peaceful sound of flowing water will be replaced by the violent thunder as the storm breaks when your primary export market is hit by an inter-governmental dispute.

The smooth ballad of easy progress will be replaced by that searching experimentation in the stuckness, as unexpected difficulties arise.

The wild excitement of rock 'n roll as you're celebrating the launch of a successful new product will be replaced by that annoying squeaky out-of-tune violin when a lapse of concentration causes the music to grind to a halt.

Neither the changes nor the challenges are in the least problematical. All result from natural human interaction.

The key Question remains …

How do we dance with these changes?
Which music do we play?

Working with a soil that has been well cultivated and fertilised,
you are far more likely to be able to transition into healthy rhythms.

For as long as human beings continue to perceive change as ...
... a "thing" to ... do or design or manage or make or lead or ...
... we will remain entangled in the waste pile of demolished constructions that have long since ceased to be fit for purpose.
We will continue to add to the waste and destruction.

We are challenged to change the narrative of change.
It is, after all, the oxygen of life. Your life.

Our ability to fully internalise the true nature of change, while working in alignment with the rhythm of our soul, will determine whether or not we are able to cultivate the truly healthy organisations, communities and societies, that actually, ought to be natural in life.

A machine is dead. It always will be, regardless of how cleverly engineers design and programme it.
A machine cannot dance, regardless of whether some imaginative software engineers are able to create robots that make movements they describe as "dance".

The true world is nature. True dance is natural movement.

The organisms and systems of organisms that are organisations, communities and societies are very much alive.

Transitioning to Action
Questions for Personal Reflection

"For me there is something
primitively soothing about this music,
and it went straight to my nervous system,
making me feel ten feet tall."
(Eric Clapton)

You have decided that the time is ripe to move, to initiate that change you've been dreaming about for a while, to leave the old habits memes and approaches behind.
Excellent. It's a good and significant start … yet merely a start.

Actually, taking the first step into the new field is what really matters. Without this step, the decision has little meaning. The idea remains a dream.

We know that Cultivating a Healthy Organisation is a community initiative.

So … regardless of your formal role in the organisation, be it a leader or a manager or employee …
… as the person with the dream, you need to get your ideas out into the world. You need other people to take you seriously, and join you because they too are convinced of the need to tread a new path. You need to be able to speak about your ideas confidently and coherently. You need to formulate your thoughts clearly.

When mechanistic ideas are reformulated, repackaged, wrapped in shiny new paper and presented as new, they may well be received positively. This is perfectly logical.
After all, they contain little or no threat to the established system.

However, when people present new ideas that challenge the established system, ideas that could actually really bring about positive changes to peoples' lives by shifting power bases and their associated rewards, they seldom attain the same positive reception.
More often than not, they will be dismissed as unrealistic dreamers or even crackpots.

Regardless …
Stay with it. Ask yourself these Questions …

Qs

Which music makes you feel ten feet tall?
Which tunes do you want to dance to … at work … at home?

How clear are your thoughts on why you are proposing this new initiative?

Discover the answers that lie deep inside you.
When we speak from the heart, from an inner conviction, the opinions of naysayers become irrelevant.

To conclude …

I offer the following Questions for personal reflection.
Perhaps they will help you clarify your thoughts before you embark on your conversations with colleagues and partners.

Give yourself time and space to respond.
When you reconsider the Questions at a later point in time, your answers may well be different.

I have challenged both individuals and groups with these Questions over many years.
Some readers will recognise those that appear in cultureQs®.

Why do you wish to embark on this initiative NOW?

Which conversation are you not having in your organisation?

Why do you need to have them?

Your best friend and a colleague who knows you well are having a conversation about you. You are not in the room.

What are they saying about you, the kind of person you are?

Why do you do what you do?

Name the three most important things for you in your working life?

"Go out from your village; don't let your village go out from you." (Afghan proverb)

Which key messages does this contain for you?

What does "success" mean for you?

"We shape clay into a pot, but it is the emptiness inside that holds whatever we want." (Lao Tzu)

Which messages does this contain?

What does "respect" mean for you?

Describe your expectations of an organisation in which you work?

Which conditions are necessary for you to take personal responsibility at work?

What really drives you, motivates you?

… Now …
Are you ready to dance?

What are you going to do … now …?

We'll leave the final words to yet another brilliant musician who, sadly, left us far too early …

Damn the rules. It's the feeling that counts.
(John Coltrane)

Acknowledgements

If I knew where the good songs came from,
I'd go there more often.
(Leonard Cohen)

The work that finally takes on a form of its own is the outcome of the myriad experiences that have woven their way into my life … and indeed, are still doing so. With the exception of a very few specific exchanges, it is practically impossible to specify which individual interaction influenced which idea, and how that particular idea influenced others. Such is the nature of our world.

I have never seen much point in dividing up my learnings into personal and professional contexts. I am, after all, one being … as you are too.

Many of these learnings are incidental, including … a conversation with two 10 year old boys on the first day of my first ever trip to India; the three Cambodians kids hustling a living at Angkor Wat; conversations with the steward of a Nepalese health centre in a remote mountain village; the aura of the Dalai Lama during a teaching in the north of India; the sight of beggars in the gold-embossed church compound in Mexico; a deep slow dialogue with a Hawaiian musician in the Californian mountains; the never-ending stream of wisdom from the tea stall in the suburb of New Delhi where I lived for a few months; exchanges with partners who have accompanied me through various phases of my life; listening to the stories of my ancestors and their friends fortunate enough to escape the tyranny of Nazism in Berlin of the 1930s; … and many more too.

Sometimes … it's the little things that make a big difference.

The professional conversations from which I have learned so much, include ... the kindly elderly HR Manager who gave me valuable feedback as I was starting my first company ... flying by the seat of my pants with neither a strategy nor a plan; the arrogant former boss who attempted to get all members of his group to behave in a manner that he had pre-defined; the Director who trusted me with her clients in my first ever job as a trainer ... a job for which I had no professional qualifications whatsoever; the Director who pulled me into my first ever Organisation Development assignment with the words, *"I don't care if you've never done anything like this before. I want you to do it.";* the leader of the excellent advanced facilitation programme I attended, opening my eyes to the true possibilities of opening space for people; the consultant who organised the worst conference I have ever attended as a speaker; the consultant who designed and pulled off one of the best and most unusual events I have ever attended as a speaker; the Professor whose work and demeanour reflects the art of the possible ... for anyone; the colleague with whom I enjoyed a game of professional ping-pong as we jointly designed a programme while by-passing the snake-pit politics of our client. There have been so many more.

To all clients who have given me the opportunity to work and learn with you ... Thank you!

There is no way I can possibly remember all the individuals who have enabled me to learn from them over the years. There are many whose names I cannot even remember; some have since passed on. I owe you all my gratitude.

I express my appreciation to Tina Hanisch, missmilla design, Munich, for working with me to create an integrated design from a collection of words and graphics.

It goes without saying that I alone remain responsible for the content. You, the reader, remain the judge of its quality and usefulness.

Sources of Inspiration

While my primary sources of inspiration have always been conversations and experiences, without books and music, life would be somewhat dull and limiting. Any collection of references is inevitably incomplete. Those presented here comprise merely a small selection of the books that have influenced my thinking and that I personally relate to. Quotes from some appear in the text.

I could easily supplement this list with the literature of Ernest Hemmingway, Milan Kundera, Hanif Kureishi, Guenter Grass, Hermann Hesse, Amos Oz, Khalil Gibran, Taiye Selasi, Chimamanda Ngozi Adichie, Haruki Murakami, Jhumpa Lahiri together with an unlimited list of further Indian authors, James Baldwin, Philip Roth, Rainer Maria Rilke, Maya Angelou. There are so many more …

And then there is music. The poetry of Leonard Cohen; and that of Bob Dylan, much of which was written decades ago … and, unfortunately, contains messages as relevant today as they were then. All that is jazz, expressed through the instruments of those who play from their soul. There is so much more …

No classical business books feature here. Some are certainly useful. I have discovered relatively few that adequately address the truly significant problems in organisation life.

Richard Bach, Jonathan Livingstone Seagull.
Gregory Bateson, Mind and Nature.
Gregory Bateson, Steps to an Ecology of Mind.
Nora Bateson, Small Arcs of Larger Circles.
Eric Berne, Games People Play.
William Bloom, Money Heart and Mind.
Pema Chödrön, When Things Fall Apart.
Mihaly Csikszentmihalyi, Flow.
Dalai Lama, Ancient Thinking Modern World.
David Bohm, On Dialogue.
David Bohm, Wholeness and the Implicate Order.
Martin Buber, I and Thou.
Susan Cain, Quiet.
Fritjof Capra, The Tao of Physics.
Fritjof Capra, Uncommon Wisdom.
Fritjof Capra & Pier Luigi Luisi, The Systems View of Life.
Joseph Campbell, The Hero with a Thousand Faces.
Viktor Frankl, Man's Search for Meaning.
Erich Fromm, The Art of Loving.
R. Buckminster Fuller, Operating Manual for Spaceship Earth.
Philip Glass, Words without Music.
Edward Hall, Beyond Culture.
Edward Hall, The Silent Language.
Thomas Harris, I'm OK- You're OK.
John Heider, The Tao of Leadership.
Benjamin Hoff, The Tao of Pooh.
William Isaacs, Dialogue.
Joseph Jaworski, Synchronicity.
C.J. Jung, The Undiscovered Self.
Byron Katie, Loving What Is.
J. Krishnamurti, Freedom from the Known.
J. Krishnamurti, On Relationship.
J. Krishnamurti, You are the World.
J. Krishnamurti, any other works ...
Alan Lightman, Searching for stars on an island in Maine.
Barry Long, Only Fear Dies.
Stephen Nachmanovitch, The Art of Is.
Humberto Maturana and Francisco Varela, The Tree of Knowledge.
Jacob Needleman, A Little Book on Love.
Jacob Needleman, Money and the Meaning of Life.
Ken Robinson, The Element.
Sogyal Rinpoche, The Tibetan Book of Living and Dying.
Eckhart Tolle, The Power of Now.
Lao Tsu, Tao Te Ching.
Paul Watzlawick, How Real is Real.
Alan Watts, The Book.
Alan Watts, The Wisdom of Insecurity.
Alan Watts, any other works ...

cultureQs®

Using Bohmian Dialogue and the power of Personal Stories as its foundation, cultureQs® represents an approach to group development that places quality relationships at its core.

"How can we accelerate the process of connecting
at a deep meaningful level,
to enable us to focus on our business questions
more quickly and effectively?"

This was the client challenge at the start of a Post-Merger Integration process in 2002 that encouraged me to begin developing cultureQs.

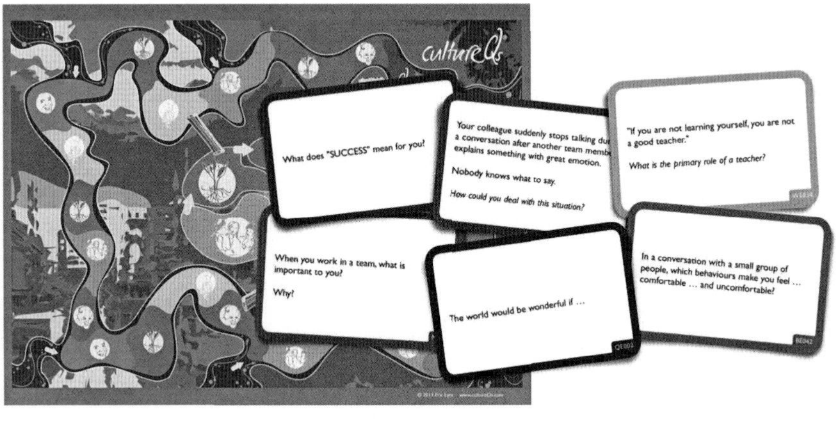

Without relationships that generates trust, so many well-intended initiatives dissolve into conflict, more often than not, because preliminary relationship development work has not been undertaken at the start. So-called "soft skills" are, in fact, fundamental business skills. cultureQs addresses this challenge directly.

cultureQs® is a unique Change and Integration Accelerator that uses powerful Questions to inspire participants to reflect on the foundations of their beliefs, attitudes and behaviours.
While engaging in deep meaningful conversations, people connect quickly, enabling the invisible borders that hinder collaboration to fall away.
Wasteful conflict is reduced; people (re-)focus on the personal and professional questions that matter; innovative potential is released and performance is enhanced.

"A trusting environment emerged very quickly."
(A client reflecting on the impact of cultureQs.)

"Every project should incorporate cultureQs right at the start."
(A former Senior Programme Manager at a global software corporation.)

Different versions of cultureQs are available in multiple languages.

Further information: https://cultureqs.com/about-cultureqs/.

The Author: Eric Lynn

Eric Lynn is recognised among clients for posing poignant Questions that permit them to diagnose their key challenges quickly. Groups appreciate his openness, human-centred holistic approach to business and low tolerance for fluff. With a clear focus on clients' real Business Questions, Eric integrates his life-long learnings with principles from Bohmian Dialogue, the Power of Personal Stories, Open Space, Appreciative Inquiry and Living Systems Thinking, to create and hold the space that enables people across hierarchies, functions and systems to align, engage and develop healthy thriving organisations and social systems.

Today, he works primarily as … a *Facilitator* for Organisation Culture Change Initiatives; an *Executive Coach* supporting leaders gain clarity and orientation in the frequently disorientating whirlwind of professional life; a *Mediator* for Executive, Management and Project Teams whose initiatives have lost their way and quickly need to get back on track.

Change and crossing cultures have been the defining features of Eric's life since childhood. As a tri-national, who was born in Canada, formally educated in the UK, has lived most of his adult life in Germany, he defies and rejects conventional categorisation. Having lived and worked on 4 continents, including 8 years in Asia, Eric has more than 40 years' worldwide experience supporting peoples' inherent desire to cooperate while connecting meaningfully. Previous occupations include, corporate communication skills trainer, teacher, economist, farm help, garbage collector, factory worker, bartender.

Eric's professional principle: Good Purposeful Work with Good People. He works in English and German, and currently lives in Germany.

www.cultureQs.com
info@cultureQs.com